Lecture Notes in Earth Sciences

Edited by Somdev Bhattacharji, Gerald M. Friedman,
Horst J. Neugebauer and Adolf Seilacher

26

Robert D. Stoll

Sediment Acoustics

Springer-Verlag

Berlin Heidelberg New York London Paris Tokyo Hong Kong

Author

Robert D. Stoll
Professor of Civil Engineering, Columbia University
Lamont-Doherty Geological Observatory of Columbia University
109 Oceanography Bldg., Palisades, New York 10964, USA

ISBN 0-387-97191-2 Springer-Verlag New York Berlin Heidelberg
ISBN 3-540-97191-2 Springer-Verlag Berlin Heidelberg New York

Printing and binding: Druckhaus Beltz, Hemsbach/Bergstr.
2132/3140-543210 – Printed on acid-free paper

To J.C.S

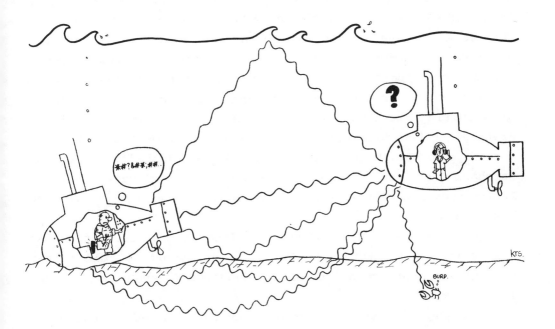

"Mancuso took a spare set of phones and plugged them
in to listen. The noise was the same. A swish, and
every forty or fifty seconds an odd, low-frequency
rumble. This close they could also hear the gurgling
and throbbing of the reactor pump. There was a sharp
sound, maybe a cook moving a pan on a metal grate.
No silent ship drill on this boat. Mancuso smiled to
himself. It was like being a cat burglar,..."

Tom Clancy, <u>The</u> <u>Hunt</u> <u>for</u> <u>Red</u> <u>October</u>,
Naval Institute Press, Annapolis, 1984.

ACKNOWLEDGEMENTS

Most of the theory and experiments described in this monograph are a result of work sponsored by the Office of Naval Research, Code 1125 OA, over a period of more than 12 years. Most recently this work has been carried out under contracts N00014-87-K-0204 and N00014-89-J-1152. During this time the continuous encouragement and support of a number of ONR project managers has allowed the author to pursue the ordered and uninterrupted development of a unified theory of sediment acoustics. I am particularly grateful to the following science officers: from Code 1125 OA - Drs. Hugo Bezdek, Mike McKisic, Peter Rogers, Ray Fitzgerald, and Marshall Orr; from code 1125 GG - Drs. Tom Pyle, Mark Odegaard, Aubrey Anderson, Gerald Morris, Jack Heacock, Joseph Kravitz, and Randy Jacobson.

I would also like to thank my colleague George Bryan, whose discussions and critiques have played an important role in our work on sediment acoustics, and my daughter Kirsten who drew the frontispiece.

In preparing this monograph I have used freely a number of modified fragments of text from our original articles in the Journal of the Acoustical Society of America and from my review article "Acoustic waves in marine sediments" in Ocean Seismo-Acoustics edited by T. Akal and J. M. Berkson, Plenum, New York, 1986. I thank the American Institute of Physics and Plenum Press for permission to cite this useage in this acknowledgement rather than in many fragmented quotations.

Lamont-Doherty Geological Observatory publication No. 4523.

CONTENTS

INTRODUCTION

Over the past 18 years the author and several colleagues have developed a mathematical model designed to predict the propagation characteristics of acoustic waves in marine sediments. The model is based on the classical work of Maurice Biot who developed a comprehensive theory for the mechanics of porous, deformable media in a series of papers spanning the time period from 1941 to 1973. Since our objective was to develop a practical working model that could be used as a guide in planning and interpreting experimental work, we began with the simplest possible form of the model and added various complexities only as they were needed to explain new variations in the data that were obtained. Thus the number of material parameters that had to be measured or assumed at any stage in the development of the model was kept to a minimum. Since the first version of the model was introduced in 1970, we have published over twenty technical papers covering various stages of its development and many papers have been published by colleagues who have utilized our work in various ways. This monograph is an attempt to summarize the development and use of the model to date.

Acoustic waves in ocean sediments may be considered as a limiting case in the more general category of mechanical waves that can propagate in fluid-saturated porous media. The general problem of wave motion in this kind of material has been studied extensively over the past thirty years by engineers, geophysicists and acousticians for a variety of reasons. In some cases, interest is focused on low-frequency waves of rather large amplitude, such as those that arise near the source of an earthquake or near a building housing heavy, vibrating machinery. At other times, the main interest is in waves of low frequency and amplitude that have traversed long distances through the sediment. In still another category, high-frequency waves that are able to resolve thin layering and other fine structural details are of interest in studying near-bottom sediments. Thus the full spread of frequency and amplitude has been studied for geological materials ranging from soft, unconsolidated sediments to rock.

Because of the almost limitless combinations of different types of sediment, stratification and structure, accurate mathematical description of the wave field produced by a particular source can be constructed only if accurate descriptions of the acoustic properties of individual components can be specified. These properties depend on the geological history of the sediment deposit, on the frequency content of the wave field and on a number of other factors that depend on the environment in situ.

A survey of the literature suggests that there are a number of parameters that play principal roles in controlling the dynamic response of saturated sediments. Of these, the following seem to be most important (not necessarily in the order listed):

(a) dynamic strain amplitude,

(b) porosity,

(c) static, intergranular stress,

(d) gradation and grain shape,

(e) material properties of individual grains,

(f) degree and kind of lithification,

(g) structure as determined by the mode of deposition.

In the following pages, we will attempt to assess the influence of most of these factors and incorporate enough detail into our model so that their influence may be studied.

Much of the data in the literature that is pertinent to the study of sediment acoustics falls into one of two broad categories of research - marine geophysics or geotechnical engineering. In geophysics the emphasis is on exploration and remote sensing, whereas in engineering the effects of machine vibrations and earthquake waves are of primary interest. The older geophysics literature contains a considerable volume of laboratory data obtained at high frequencies (i.e.,greater than about 1 kHz), whereas the engineering studies, which often employ the resonant column technique to measure the dynamic moduli and damping, are largely concentrated in the frequency range from 20 Hz to about

200 Hz. In fact it was the seemingly anomalous results obtained by these two different groups that was the initial motivation in our search for a unifying theory (Stoll and Bryan, 1970).

We have drawn heavily on the technical literature from both engineering and geophysics and have attempted to sort out the experimental data appropriate to the range of strain amplitudes that would be anticipated when designing a geoacoustic model of the ocean bottom. Unfortunately, since much of the engineering data were obtained at fairly high strain levels (appropriate to regions not too far from a vibrating source), they cannot be used without modification because of the highly nonlinear effects of amplitude on modulus and damping. In dealing with acoustic waves in marine sediments we often have weak sources or signals far from the source so that a linear approximation which is independent of amplitude is often adequate. Numerous experiments have shown that a linear approximation is meaningful as long as the strain amplitude is less than about 10^{-6}. When this criterion is met, models which predict velocity and attenuation can be greatly simplified. In this monograph we will concentrate on modeling waves of low amplitude and will attempt to point out the effects of nonlinear behavior as they affect the predicted behavior.

In Chapter 1 elements of the basic Biot theory are presented and used to develop a mathematical model for water-saturated sediments. In this chapter we consider only the isotropic case and show how various forms of damping may be introduced by choosing the moduli of the skeletal frame to be complex. In Chapter 2 some parametric studies are presented in order to help in visualizing the effects of variations in some of the 13 or more parameters that are needed to use the model. In this chapter we consider only basic trends and no effort is made to compare the predictions of the model with real data. In addition to the parametric studies of model parameters, we also examine the differences in response between the Biot model and other elastic or viscoelastic models when a boundary is encountered. Partitioning of energy and the resulting reflection and refraction coefficients are discussed. Chapter 3 presents a brief summary of the mechanics of granular media, and the effects of various quasistatic loading histories are shown to introduce anisotropy

into the acoustic response. The work described in this chapter
sheds light on how geostatic stress influences wave velocity and
attenuation, as well as the degree of anisotropy that may be
expected. In Chapter 4 we look at experimental methods that may
be used to study dynamic moduli in saturated sediments. Experimental results obtained by the author and other investigators
are used to establish the range of values to be expected for the
various parameters used in the Biot and Gassmann equations. The
choice of realistic parameters for the model is discussed further
in Chapter 5, and the model is generalized by making the complex
moduli of the skeletal frame frequency dependent using a simple
viscoelastic model fitted to the data displayed in Chapter 4.
Finally, in Chapter 6, we compare the predictions of the model
with the results of several recent field experiments performed
in areas where the sediment properties were well-documented from
prior studies. This last chapter concludes with a detailed
discussion on how to choose realistic input parameters for pre-
dictive geoacoustic modeling.

CHAPTER 1
THE BIOT THEORY

INTRODUCTION

Starting with a paper on consolidation of porous, elastic material in 1941 (Biot, 1941), Biot developed a comprehensive theory for the static and dynamic response of porous materials containing compressible fluid. He considered both low- and high-frequency behavior (Biot, 1956a, 1956b) and included the possibility of viscoelastic or viscodynamic response in various components of his model (Biot, 1962a, 1962b).

In the course of developing and generalizing the theory, Biot introduced several changes of notation and a number of general-izations so that some effort is necessary in order to extract the form most suitable for a particular application. For this reason an abbreviated derivation leading to one form of his equations is given below. This derivation helps to identify the variables that are used and to visualize how the response of the sediment is modeled in a mathematical way. For more rigorous and complete derivations, the reader is referred to Biot's original papers, particularly Biot (1962a) and to a paper by Geertsma and Smit (1961).

Biot's theory predicts that, in the absence of boundaries, three kinds of body waves, two dilatational and one rotational, may exist in a fluid-saturated, porous medium. One of the dilatational waves, which is called the "first kind," and the shear wave are similar to waves found in ordinary elastic media. In these waves the motions of the skeletal frame and the inter-stitial fluid are nearly in phase and the attenuation owing to viscous losses is relatively small. In contrast, the dilatational wave of the "second kind" is highly attenuated and the frame and fluid components are moving largely out of phase. Waves of the first kind are sometimes called "jostling" waves and waves of the second kind "sloshing" waves. Compressional waves of the second kind become very important in acoustical problems involving very compressible pore fluids such as air, whereas for geophysical work in water-saturated sediments, waves of the first kind are of principal interest. One exception to this is the case of very gassy sediments where the effective compressibility of the pore fluid is greatly reduced by the presence of free or dissolved gases.

BASIC EQUATIONS

To obtain equations governing the propagation of dilatational waves, we first consider the case of a plane wave in a porous, elastic medium that is filled with fluid. The model is then generalized to include the inelasticity of the skeletal frame and the frequency dependence of various viscous losses to yield a realistic model of naturally occurring sediments.

If $\vec{u}$ is a vector function giving the displacement of points in the skeletal frame and $\vec{U}$ a vector function giving the displacement of the fluid, then the volume of fluid that has flowed in or out of an element of volume attached to the frame or the "increment of fluid content" is

$$\zeta = \beta \operatorname{div}(\vec{u} - \vec{U}) \qquad (1.1)$$

where β is the ratio of the volume of the pores to the total volume of the element (porosity). For small strains the dilatation or volumetric strain of the element attached to the frame may be written as

$$e = e_x + e_y + e_z = \operatorname{div} \vec{u} \qquad (1.2)$$

where e_x, e_y and e_z are components of small compressional or extensional strain in a Cartesian coordinate system. If the porous frame is elastic (non-dissipative), the strain energy, W, of the system depends on the strain components and the increment of fluid content

$$W = W(e_x, e_y, e_z, \gamma_x, \gamma_y, \gamma_z, \zeta) \qquad (1.3)$$

where γ_x, γ_y, and γ_z are the components of shear strain. For an isotropic, linear material, W is a quadratic function of the invariants of strain, I_1 and I_2, and the increment of fluid content, ζ

$$W = C_1 I_1^2 + C_2 I_2 + C_3 I_1 \zeta + C_4 \zeta^2$$

$$I_1 = e_x + e_y + e_z = e \tag{1.4}$$

$$I_2 = e_x e_y + e_y e_z + e_z e_x - 1/4(\gamma_x^2 + \gamma_y^2 + \gamma_z^2)$$

The constants C_1, C_2, C_3, and C_4 may be identified with one set used by Biot (H, C, M, and μ) by writing Eq. (1.4) as

$$W = \frac{H}{2}e^2 - 2\mu I_2 - Ce\zeta + \frac{M}{2}\zeta^2 \tag{1.5}$$

Considering the total stresses on the element of volume attached to the frame, and the pressure in the pore fluid, p_f, a set of stress-strain relationships may be obtained from the strain energy, Eq. (1.5), by differentiation so that

$$\tau_{xx} = \partial W / \partial e_x, \qquad \tau_{xy} = \partial W / \partial \gamma_z, \qquad p_f = \partial W / \partial \zeta$$

etc. The resulting stress-strain relations are

$$\tau_{xx} = He - 2\mu(e_y + e_z) - C\zeta$$

$$\tau_{yy} = He - 2\mu(e_z + e_x) - C\zeta$$

$$\tau_{zz} = He - 2\mu(e_x + e_y) - C\zeta$$

$$\tau_{xy} = \mu\gamma_z \tag{1.6}$$

$$\tau_{yz} = \mu\gamma_x$$

$$\tau_{zx} = \mu\gamma_y$$

$$p_f = M\zeta - Ce$$

From these equations it is clear that μ is the shear modulus. However, in order to understand the significance of the constants H, C, and M it is helpful to visualize two idealized quasistatic tests involving isotropic loading.

In one kind of test, termed a "jacketed" test (Biot and Willis, 1957), the saturated porous medium (shown as a granular medium in Fig. 1.1) is placed in an impervious, flexible bag and loaded by an external pressure. The interstitial fluid in the sample is free to flow out of the bag via a tube so that the fluid pressure remains unchanged during slow loading. In the other test, called an "unjacketed" test, an uncased sample is completely immersed in fluid which is subsequently pressurized from an external source. If p' is the externally applied isotropic pressure in both cases, then

$$\tau_{xx} = \tau_{yy} = \tau_{zz} = -p'$$

$$\tau_{xy} = \tau_{yz} = \tau_{zx} = 0$$

and adding the first, second and third of Eqs. (1.6),

$$-p' = (H - 4\mu/3)e - C\zeta \qquad (1.7)$$

For the "jacketed" test, p_f is zero so that the bulk modulus of the free-draining, porous frame, K_b, is

$$K_b = -p'/e = H - 4\mu/3 - C^2/M \qquad (1.8)$$

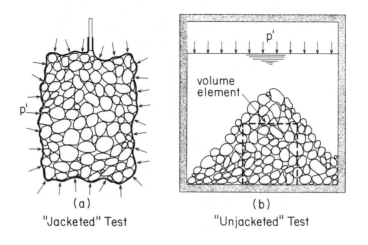

(a)   (b)
"Jacketed" Test   "Unjacketed" Test

Fig. 1.1. Isotropic tests to determine moduli
(Stoll, 1986)

from the last of Eqs. (1.6) and Eq. (1.7). In a practical test the dilatation of the sample, e, can be obtained by measuring the quantity of fluid expelled from the sample provided the effect of membrane penetration on the outer surface of the specimen is negligible (which is sometimes not the case). This type of drained, isotropic loading is utilized in several standard tests used to determine the engineering properties of soil (Bishop and Henkel, 1957). Unfortunately, the stress-strain curve that is routinely obtained during monotonic loading does not contain the information necessary to determine the modulus that results from cyclic loading at very small amplitude. For this reason the bulk modulus of the skeletal frame is often deduced from other kinds of experimental data such as wave speed or resonant frequency. Methods for measuring or estimating K_b will be discussed in Chaps. 3, 4 and 5.

In the unjacketed test, the pressure in the pore fluid equals the applied isotropic pressure, and we may derive two measures of compliance, δ, the "unjacketed compressibility,"

$$\delta = -e/p' = \frac{1-C/M}{H-4\mu/3-C^2/M} \tag{1.9}$$

and, γ, the "coefficient of fluid content,"

$$\gamma = \zeta/p' = \frac{H-4\mu/3-C}{(H-4\mu/3-C^2/M)M} \tag{1.10}$$

utilizing the last of Eqs. (1.6) and Eq. (1.7) in both cases. If the ratio of pore volume to solid volume remains constant during unjacketed loading (i.e., constant porosity), δ equals the compressibility of the solid material composing the porous frame and γ may be expressed in terms of δ, β and the compressibility of the pore fluid. In terms of the reciprocals of compressibility (bulk moduli)

$$\gamma = \beta(1/K_f - 1/K_r) \tag{1.11}$$

and

$$\delta = 1/K_r \tag{1.12}$$

where K_f is the bulk modulus of the fluid and K_r is the bulk modulus of the solid material composing the porous frame (the bulk modulus of the individual particles in the case of granular media). While Eqs. (1.11) and (1.12) are strictly true for materials where the porous frame is isotropic, homogeneous and linear, they are also reasonable for cases where the frame does not exactly satisfy these conditions but still undergoes the same volumetric strain as the pores on the average. In many naturally occuring sediments this is not an unreasonable approximation; however, it should be realized that Eqs. (1.9) and (1.10) define the natural compliances or moduli that must be used if the theory is to be applied in a general case. Unfortunately, the coefficient of fluid content is very difficult to measure experimentally and therefore Eqs. (1.11) and (1.12) are generally used to establish the relationship between the parameters of the Biot theory and more familiar moduli that are easily measured, even though this

may not always be justified. In fact, some authors have been led
to inconsistent theoretical results by failing to account for the
approximation that is implicit when using these equations (see
Berryman, 1981; Korringa, 1981; and Berryman 1982). Since our
objective is to formulate a simple, usable model, we accept the
approximation necessary to use these equations in order to
facilitate the use of physical parameters that are relatively
easy to obtain and catalogue.

Using Eqs. (1.8) through (1.12), Biot's coefficients may be
written in terms of the bulk moduli of the porous frame, the pore
fluid and the solid material of the frame or discrete particles

$$H = \frac{(K_r - K_b)^2}{D - K_b} + K_b + 4\mu/3$$

$$C = \frac{K_r(K_r - K_b)}{D - K_b} \tag{1.13}$$

$$M = \frac{K_r^2}{D - K_b}$$

where

$$D = K_r(1 + \beta(K_r/K_f - 1))$$

In the above H - 4μ/3 is also the "effective" bulk modulus of a
saturated, porous material in which the fluid is restricted against
movement relative to the skeletal frame (i.e., very low permeability or occluded pores). Using the first of Eqs. (1.13) a
little algebra shows that H - 4μ/3 is the same as the effective
bulk modulus derived by Gassmann (1951), so that the Gassmann
equations may be considered as a special case of the Biot equations
applicable when the increment of fluid content, ζ, is zero.
Moreover, in very soft sediments or suspensions, where both K_b
and μ become vanishingly small, we can recover the Wood equation
for the compressibility of a water-sediment mixture by letting
$K_b = \mu = 0$ in the the the expression for H-4μ/3 giving

$$c_m = \frac{1}{K_r}(1 - \beta) + \frac{1}{K_f}\beta$$

where c_m is the compressibility of the mixture (Wood, 1912).

Having established a set of constitutive equations and the meanings of the various parameters, equations can now be written for the motion of an element of volume attached to the skeletal frame and for fluid moving into or out of the element. To simplify the derivation, we consider one-dimensional motion in the x direction. The stress-equation of motion for the volume attached to the frame is

$$\frac{\partial \tau_{xx}}{\partial x} = \frac{\partial^2}{\partial t^2}[\beta \rho_f U_x + (1-\beta)\rho_r u_x] \qquad (1.14)$$

$$= \frac{\partial^2}{\partial t^2}[\beta \rho_f u_x + (1-\beta)\rho_r u_x - \beta \rho_f (u_x - U_x)]$$

where ρ_f is the mass density of the pore fluid and ρ_r is the density of the solid material (rock) composing the frame (density of individual grains for granular material). Differentiating with respect to x and substituting for τ_{xx} from Eq. (1.6) we obtain the one-dimensional form of one of the equations given by Biot (Eq. (6) of Biot, 1962a). His equation is

$$\nabla^2(He - C\zeta) = \frac{\partial^2}{\partial t^2}(\rho e - \rho_f \zeta) \qquad (1.15)$$

where ρ is the total density of the saturated medium.

The second equation, which describes the motion of the fluid relative to the frame, is

$$\beta \frac{\partial p_f}{\partial x} = \frac{\partial^2}{\partial t^2}[\beta \rho_f U_x] + \frac{\beta \eta}{k}\frac{\partial}{\partial t}[\beta(U_x - u_x)] \qquad (1.16)$$

or

$$\frac{\partial p_f}{\partial x} = \frac{\partial^2}{\partial t^2}\left[\rho_f u_x - \frac{\rho_f}{\beta}(\beta(u_x - U_x))\right] - \frac{\eta}{k}\frac{\partial}{\partial t}[\beta(u_x - U_x)].$$

The last term on the right-hand side of this equation gives the viscous resistance to flow which depends on η, the viscosity of the pore fluid, and k, the coefficient of permeability. By differentiating Eq. (1.16) with respect to x and substituting for p_f from Eq. (1.6) we obtain the one-dimensional form of the second equation given by Biot,

$$\nabla^2(Ce - M\zeta) = \frac{\partial^2}{\partial t^2}(\rho_f e - m\zeta) - \frac{\eta}{k}\frac{\partial \zeta}{\partial t} \qquad (1.17)$$

In Eq. (1.17), an apparent mass m, greater than ρ_f/β, has been substituted for ρ_f/β in the part of the inertial term corresponding to the increment of fluid flow. This has been done to account for the fact that not all of the pore fluid moves in the direction of the macroscopic pressure gradient because of the tortuous, multi-directional nature of the pores. As a result less fluid flows in or out of an element for a given acceleration than if all the pores were uniform and parallel to the gradient. The parameter m may be written as

$$m = \alpha \rho_f/\beta, \qquad \alpha \geq 1 \tag{1.18}$$

For uniform, cylindrical pores with axes parallel to the gradient, α would equal 1, while for a random system of uniform pores with all possible orientations, the theoretical value of α is 3. In real granular materials it is extremely difficult to calculate α from theory so that it must be considered one of the variables to be determined from experiments.

SOLUTION OF EQUATIONS

Equations (1.15) and (1.17) are a pair of coupled differential equations that determine the dilatational motion of a saturated porous medium with a linear elastic frame and a constant ratio of fluid flow to pressure gradient (Poiseuille flow). Solving these equations leads to a relationship between attenuation and frequency such as shown by the broken curve of Fig. 1.2.; however, it is clear from this figure that the model is not adequate to predict the behavior of any real sediment at this stage of development. In fact, three major modifications are required to accomplish this. First, the viscous resistance to fluid flow must be made frequency dependent to correct for the deviation from Poiseuille flow that occurs at all but very low frequencies. Second, the inelastic nature of the skeletal frame owing to frictional losses and relaxation of intergranular bonds must be accounted for and third, local viscous losses occurring in the fluid as a result of local motion near the intergranular contacts must be considered. This local fluid motion is similar in nature to "squeeze film motion" that is well known in the theory of lubrication and was suggested by Biot (1962b) as a mechanism that could cause additional dissipation of energy. Squeeze film motion causes frequency dependent forces between particles so that the overall behavior of the skeletal frame is also frequency dependent whenever the effect of these forces becomes significant.

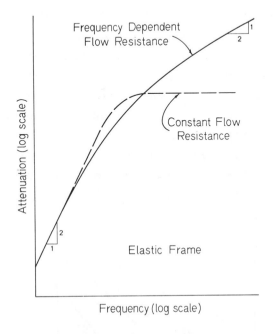

Fig. 1.2. Attenuation versus frequency for a linear elastic frame (Stoll, 1974).

In order to incorporate the frequency dependence of viscous flow, Biot derived a complex correction factor to be applied to the fluid viscosity by considering the actual microvelocity field that exists within the pore channels of an idealized porous solid. The problem of oscillatory motion in a closed channel is quite well known, having been solved as early as 1868 by Kirchoff. Biot's solution is written in such a way that the factor gives the ratio of the friction force exerted by the fluid on the frame to the average relative velocity for oscillatory motion. Hence, in view of Eq. (1.16) or (1.17), η/k may be replaced by $F\eta/k$ in the frequency domain, where

$$F(\kappa) = F_r(\kappa) + iF_i(\kappa) = \frac{1}{4}\frac{\kappa T(\kappa)}{1 - 2T(\kappa)/i\kappa},$$

$$T(\kappa) = \frac{ber'(\kappa) + ibei'(\kappa)}{ber(\kappa) + ibei(\kappa)}, \qquad\qquad (1.19)$$

$$\kappa = a(\omega\rho_f/\eta)^{1/2}$$

The functions ber(κ) and bei(κ) are the real and imaginary parts of the Kelvin function, ω is angular frequency, and a is a parameter, with the dimension of length, that depends on the size and shape of the pores. F(κ) approaches unity for very low frequencies, thus resulting in the same equation as when Poiseuille flow is assumed. Like a, the parameter a cannot be derived theoretically for real sediments so it must be deduced from experimental results. The functional variation of the real and imaginary parts of F(κ) is shown in Fig. 1.3. A BASIC computer program containing an algorithm for calculating F is given as a part of Appendix A.

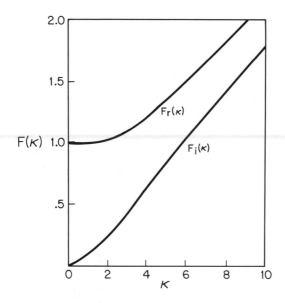

Fig. 1.3. Real and imaginary parts of frequency correction function.

The solution on which Biot's complex correction factor is based is valid only for frequencies where the wavelength is large compared to the pore size. For sands, this puts the upper limit on frequencies at about 10^5 to 10^6 Hz, which is high enough to cover the frequency range of interest in most applications. When the complex correction factor is incorporated into Eq. (1.17), by replacing η with ηF, a relationship between frequency and attenuation such as shown by the solid curve in Fig. 1.2 is obtained.

In order to incorporate the inelasticity of the skeletal frame as well as the effects of local viscous damping (due to squeeze film motion) into the model, H, M, C, and μ in Eqs. (1.15) and (1.17) are now redefined as operators that may be linear viscoelastic or "slightly" nonlinear. Moreover, for the conditions in most sediments it is reasonable to consider K_r and K_f, the bulk moduli of the individual particles and the fluid, respectively, as elastic constants and concentrate the inelastic and frequency dependent effects in the operators K_b and μ which describe the response of the skeletal frame in a water environment.

To obtain a frequency equation, solutions of Eqs. (1.15) and (1.17) of the form

$$e = B_1 \exp(i(\omega t - lx)) \tag{1.20}$$

and

$$\zeta = B_2 \exp(i(\omega t - lx))$$

are considered, with $l = l_r + il_i$, the constants H, C, M and μ replaced by the appropriate operators, and η replaced by $F(\kappa)\eta$. Upon transformation to the frequency domain the following equation results

$$\begin{vmatrix} \overline{H} l^2 - \rho \omega^2 & \rho_f \omega^2 - \overline{C} l^2 \\ \overline{C} l^2 - \rho_f \omega^2 & m\omega^2 - \overline{M} l^2 - \dfrac{i\omega F \eta}{k} \end{vmatrix} = 0 \tag{1.21}$$

where, in general, $\overline{H}, \overline{C}$ and $\overline{M}$ are complex functions of frequency derivable by Laplace transformation in the case of linear viscoelastic operators or by the method of Kryloff and Bogoliuboff (1947) for the case of slightly nonlinear operators. The resulting complex moduli are

$$\overline{H} = \frac{(K_r - \overline{K}_b)^2}{D - \overline{K}_b} + \overline{K}_b + 4\overline{\mu}/3$$

$$\overline{C} = \frac{K_r(K_r - \overline{K}_b)}{D - \overline{K}_b} \tag{1.22}$$

$$\overline{M} = \frac{K_r^2}{D - \overline{K}_b}$$

where

$$\overline{K}_b = K_{br}(\omega) + iK_{bi}(\omega)$$

and

$$\overline{\mu} = \mu_r(\omega) + i\mu_i(\omega)$$

The roots of Eq. (1.21) give the attenuation l_i and the phase velocity ω/l_r as functions of frequency for the first and second kinds of dilatational wave.

When granular materials such as sand, sandstone and some silts are tested dry with oscillatory loading of very small amplitude, the principal mechanism of energy loss is the friction which governs the minute amounts of slip occurring within the contact area between particles (Mindlin, 1949). This slip is confined to a small, annular region of the contact area and does not result in any gross sliding between particles. As a result, the damping observed in dry granular materials is usually very small (e.g. typically $1/Q$ will be of the order of 10^{-2} in sands and sandstones) and essentially independent of frequency. When these materials are saturated with water, the viscous losses in the fluid must be added to the effects of friction when considering the overall energy loss.

The frequency range over which the viscous losses become important depends on permeability, grain size, and the geometry of the interstitial pores. In coarse granular materials such as sands, overall motion of the fluid field relative to the skeletal frame can produce enough viscous damping to dominate the overall attenuation starting at frequencies of only a few Hz. In these coarse materials interparticle forces (and damping) from squeeze film motion are small enough to disregard. On the other hand in very fine materials, the overall fluid motion is negligibly small until very high frequencies are reached; however, because of the

small particle size, interparticle forces from the local fluid
motion cause a dominant amount of frequency-dependent damping.
Thus the simplest functional form of K_{br}, K_{bi}, μ_r, and μ_i that can
be used in a particular case depends on the material and frequency
range of interest. In early work with sands (Stoll, 1974, 1978,
1979) the choice of $\overline{K}$ and $\overline{\mu}$ to be complex constants led to good
agreement with experimental data. In more recent work with finer
sediment, (Stoll, 1984, 1990) it was necessary to experimentally
measure the frequency dependence of the response after which a
linear viscoelastic model was used to derive an appropriate
functional form for the moduli. Determination of appropriate
values for the moduli will be discussed in later chapters.

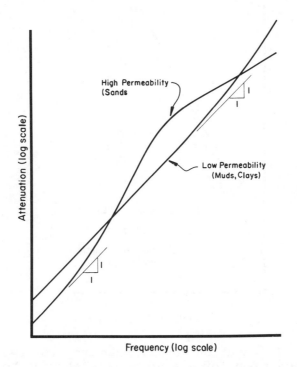

Fig. 1.4. Typical curves of attenuation versus
frequency for coarse and fine sediments based on
the proposed model (Stoll, 1977).

As described above, the way in which attenuation varies with
frequency depends on the dominance of one or the other of the
modes of energy dissipation that are built into the model - viscous
losses in the fluid as it moves relative to the frame or losses
associated with the skeletal frame characterized by the complex

moduli $\overline{\mu}$ and $\overline{K}_b$. Fig. 1.4 shows two fundamentally different kinds
of response predicted by the model. For very fine materials of
low permeability, the losses in the skeletal frame predominate,
and the attenuation is a slowly varying function of frequency.
For coarser material of higher permeability, the viscous losses
in the pore-water tend to dominate as the frequency increases
causing a dramatic increase in attenuation.

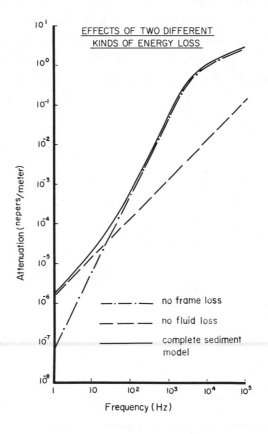

Fig. 1.5. Attenuation versus frequency for two
extreme cases and for a typical sand showing how
frame losses and fluid losses combine to control
the overall response (Stoll, 1977).

The relative effect of the two factors is shown in Fig. 1.5.
The solid curve in Fig. 1.5 is what the model predicts for a
typical sand where the complex moduli are assumed to be constants.
When the permeability is arbitrarily made very small so that there
is virtually no movement of pore fluid relative to the frame, one

obtains the dashed curve. On the other hand, if normal fluid mobility (permeability) is assumed and the frame is made elastic by letting $\bar{\mu}$ and $\overline{K}_b$ become real elastic constants, the other broken curve in the figure is obtained. It is easy to see how superposition of the two broken curves leads to a result which approximates the real, coupled behavior depicted by the solid curve.

Equations governing the propagation of shear waves may also be derived from Eq. (1.6), the equation of fluid flow and the stress equations of motion. The resulting equations of motion are

$$\mu\nabla^2\vec{\omega} = \frac{\partial^2}{\partial t^2}(\rho\vec{\omega} - \rho_f\vec{\theta}) \tag{1.23}$$

$$\eta/k\frac{\partial\vec{\theta}}{\partial t} = \frac{\partial^2}{\partial t^2}(\rho_f\vec{\omega} - m\vec{\theta})$$

where

$$\vec{\omega} = \text{curl}\,\vec{u}$$

and

$$\vec{\theta} = \beta\,\text{curl}(\vec{u} - \vec{U})$$

Only one kind of shear wave can propagate and its characteristics are determined by the frequency equation

$$\begin{vmatrix} \rho\omega^2 - \bar{\mu}l^2 & \rho_f\omega^2 \\ \rho_f\omega^2 & m\omega^2 - \dfrac{i\omega F\eta}{k} \end{vmatrix} = 0 \tag{1.24}$$

Again η has been replaced by $F(\kappa)\eta$ and μ by the complex, frequency-dependent modulus $\bar{\mu} = \mu_r + i\mu_i$. Note that even though there is no volume change, there is still relative motion between the fluid and the frame so that both losses in the frame and losses in fluid affect the result. The fluid is in effect "dragged" with the frame as the volume constant shearing deformation propagates as a wave.

Table 1. is a list of the various physical parameters that must be input to the model.

In subsequent chapters, we will give many examples of the solution of Eqs. (1.21) and (1.24) as the above parameters are varied.

TABLE - 1.1

BASIC PHYSICAL PARAMETERS

(frequency-independent variables)

Porosity	β
Mass density of grains	ρ_r
Mass density of pore fluid	ρ_f
Bulk modulus of sediment grains	K_r
Bulk modulus of pore fluid	K_f

(variables affecting global fluid motion)

Permeability	k
Viscosity of pore fluid	η
Pore-size parameter	a
Structure factor	α

(variables controlling frequency-dependent response of frame)

Shear modulus of skeletal frame	$\bar{\mu} = \mu_r(\omega) + i\mu_i(\omega)$
Bulk modulus of skeletal frame	$\bar{K}_b = K_{br}(\omega) + iK_{bi}(\omega)$

MODEL PARAMETERS - PARAMETRIC STUDIES

In order to use Eqs. 1.21 and 1.24 to determine the velocity and attenuation in real sediments it is necessary to choose realistic values for 13 or more different parameters. Several of the choices are straightforward and reasonable values may be obtained from standard references; however, a few of the parameters need careful consideration in order to produce meaningful predictions. We will consider these parameters in two categories - the complex moduli of the skeletal frame $\overline{K}_b$ and $\overline{\mu}$ and the three parameters (a, m, and k) that control the overall mobility of the fluid component and its interaction with the skeletal frame.

In the examples to follow we assume that the wave amplitude and the logarithmic decrement are small so that the different measures of damping may be related in a particularly simple way. For example the quality factor Q, the logarithmic decrement δ (for both traveling waves and stationary vibrations), the attenuation coefficient $\alpha = l_i$, and the ratio of the imaginary and real parts of the appropriate complex modulus, $\overline{M} = M_r + iM_i$ are related approximately by the expression

$$\frac{1}{Q} = \frac{\delta}{\pi} = \frac{\alpha V}{\pi f} = \frac{M_i}{M_r} = \tan\theta_L \tag{2.1}$$

where V is phase velocity, f is frequency and θ_L is the phase difference between stress and strain under harmonic loading.

In the case of dry granular materials such as sands, silts and sandstones, the principal mechanism of energy loss is friction and typically, 1/Q will be of the order of 10^2 or slightly larger. However, when these granular materials are located in a water environment, 1/Q increases and becomes frequency dependent at all but the lowest frequencies because of the additional energy losses attributable to the viscosity of the fluid component. The effects of viscosity manifest themselves in two different ways. There is an overall motion of the fluid field relative to the skeletal frame of the sediment and local motion in the vicinity

of intergranular contacts. The effect of the overall motion of
the fluid is built into the basic Biot theory and controlled by
the parameters (a,m and k) or any equivalent set devised to couple
the motion of the fluid and solid components. The effects of
viscosity in local fluid flow must be built into the complex
moduli $\overline{K}_b$ and $\overline{\mu}$. The frequency-dependent damping produced by
local fluid motion becomes larger as the particle size and the
relative distance between contiguous surfaces near contacts become
smaller. Thus, this mechanism of loss tends to become dominant
in fine-grained sediments such as silt and clay where the overall
mobility of the fluid is low.

In modeling the coarser granular materials we have generally
chosen the moduli $\overline{K}_b$ and $\overline{\mu}$ to be complex constants with the ratio
of the imaginary to real parts chosen on the basis of experimental
measurements on dry specimens or water-saturated specimens tested
quasistatically. The real parts of $\overline{K}_b$ and $\overline{\mu}$ may be determined
from the resonant frequencies in resonant column experiments or
from measurements of wave velocity in the lab or field. In
choosing values for $\overline{K}_b$ and $\overline{\mu}$ it is important to remember that
these must be determined for drained conditions and for very small
amplitudes of motion (strains < 10^{-6}).

We will discuss the choice of complex moduli and other
parameters in more detail; however, at this point it is instructive
to see what Eqs. 1.21 and 1.24 predict when one or more of the
parameters is varied in a systematic way. In the first group of
figures we have assigned a single set of values to all parameters
except a (the pore size parameter) and k (the coefficient of
permeability). In Figs. 2.1 and 2.2, k and a have been varied
systematically in such a way that k/a^2 = constant, in analogy
with the predictions of the Kozeny-Carman equation (Carman, 1956).
Fig. 2.1 shows the logarithmic decrement predicted for shear waves
and Fig. 2.2 the decrement predicted for p-waves of the first
kind. The low-frequency end of all of the curves in Fig. 2.1
approaches asymptotically to the value of δ_μ that was input to
the model as the decrement for dry sediment or quasistatic loading.
Thus, as the frequency approaches zero, the log decrement is
determined by the expression $\delta_\mu = \pi \mu_i / \mu_r$ which is a measure of
frictional losses due to shearing motion. As the frequency is

increased, the decrement increases to some peak value after which
it decays in a manner very similar to a typical viscoelastic
relaxation curve. As the permeability decreases the peaks shift
to higher and higher frequencies. The curves for p-waves of the
first kind shown in Fig. 2.2 are similar to those of Fig. 2.1
except for one major difference - the asymptotic value of the log
decrement at the low-frequency end of each curve is less than
either of the values of log decrement input to the model by a
factor of about 30! Thus, by saturating the sediment with water,
we end up with a material with much lower compressibility but
with the same basic mechanism for dissipating energy by friction;
the net result is a much lower overall logarithmic decrement at
low frequencies for any motion involving volume change.

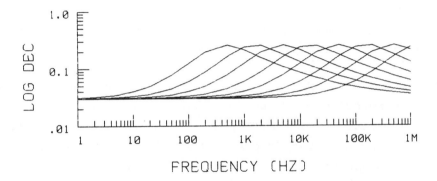

Fig. 2.1. Logarithmic decrement for shear waves in
water-saturated sand - k/α^2 = constant, permeability
decreasing from left to right (Stoll, 1986).

In order to calculate the attenuation exponent α, it is
necessary to know both the logarithmic decrement and the phase
velocity. Because of damping, there is a small amount of dispersion
in p-waves of the first kind; the pattern of this dispersion is
shown by the curves of Fig. 2.3 where it can be seen that the
velocity increase occurs over a rather narrow band of frequency
for each curve . As the fluid mobility decreases because of lower
permeability, this band shifts to higher and higher frequencies
in the same manner as the peaks of the damping curves in Fig.
2.2.

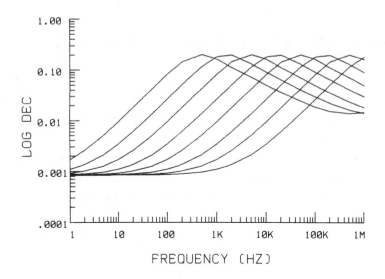

Fig. 2.2. Logarithmic decrement for p-waves in water-
saturated sand - k/a^2= constant (Stoll, 1986).

Fig. 2.4 shows the attenuation of p-waves of the first kind
that corresponds to the logarithmic decrements shown in Fig. 2.2
and the velocities shown in Fig. 2.3. The highly nonlinear form
of this family of curves suggests that there is no simple power
law of the form $\alpha = K f^n$ (f is frequency; K and n are constants)
that lends itself to a fit of the data over an extended range of
frequencies. This kind of power law has often been fitted to
experimental data obtained over a narrow range of frequencies and
the values of K and n cataloged for future use. It can be seen
from this figure that once the values of K and n are taken out
of the context of a complete description of the original experiment,
they may be very misleading.

    The shape of the curves shown in Fig. 2.4 depends on both
the pore size parameter a, and the permeability k. As an example
of the effect of different combinations of these parameters, the
value of k has been fixed and four different values of a were
assumed in order to generate the curves shown in Fig. 2.5. From
this figure it is clear that a and k are related in a rather

subtle way so they cannot be chosen independently in an arbitrary manner if one hopes to obtain a good fit to a particular set of experimental data.

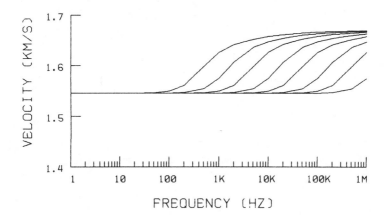

Fig. 2.3. Velocity of p-waves as a function of frequency (Stoll, 1986).

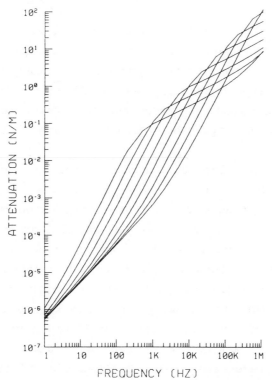

Fig. 2.4. Attenuation of p-waves corresponding to logarithmic decrements shown in Fig. 2.2 and velocities shown in Fig. 2.3 (Stoll, 1986).

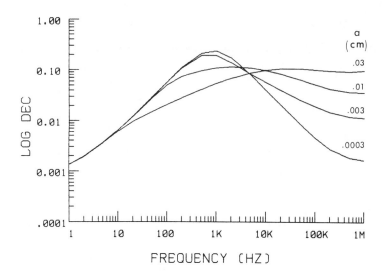

Fig. 2.5. Logarithmic decrement for p-waves - k fixed,
a varying (Stoll, 1986).

A variation in decrement as shown in Fig. 2.5 causes a rather
marked change in the shape of the corresponding attenuation curves,
Fig. 2.6. The influence of this parameter is most evident at the
higher frequencies and so experimental data from this frequency
range are very useful in choosing the best value for a. We will
compare curves of the type shown in this figure with experimental
data in Chap. 5.

In order to predict the p-wave response in finer sediments
where the permeability is very low, it is necessary to cast the
moduli $\overline{K}_b$ and $\overline{\mu}$ as frequency-dependent complex quantities either
directly on the basis of experimental results or on the basis of
some viscoelastic model. We have recently completed a set of
torsional resonant column experiments on a micaceous silt material
of fairly low permeability (approximately 2.5×10^{-10} cm^2) covering
the frequency range from 1 Hz to about 1.5 kHz (Stoll 1986). The
decrement measured in these experiments is shown in Fig. 2.7.
To determine the logarithmic decrement at low frequencies, we
measured the difference in phase between the driving torque and
the resulting torsional motion using a very sensitive capacitive
probe. At high frequencies the decrement was determined by

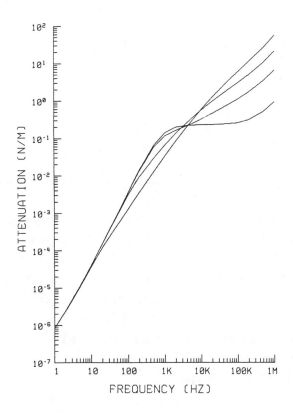

Fig. 2.6. Attenuation versus frequency corresponding
to the damping curves shown in Fig. 2.5.

measuring the decay of motion after the specimen was driven to
one of several different resonant frequencies. These experiments
will be discussed in more detail in Chap. 4.

If we now take the low-frequency asymptote of Fig. 2.7 to
determine a value of δ_μ representing frictional damping under
quasistatic motion and then choose $\overline{K}_b$ and $\overline{\mu}$ to be complex constants,
as was done previously for the example involving sands (Figs.2.1
through 2.5), we obtain the curves labeled "constant complex
modulus" shown in Fig. 2.8. Clearly the permeability is so low
that form of viscous damping included in the basic Biot theory
only begins to show an effect at the extreme high frequency end

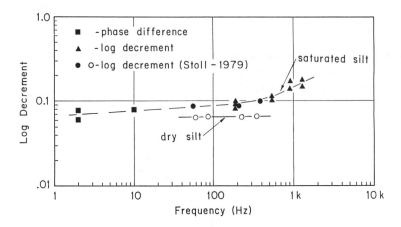

Fig. 2.7. Experimentally determined logarithmic decrement for silt (Stoll, 1985).

of the curve. In order to incorporate the frequency dependence
of damping that is obvious in Fig. 2.7 we now determine the
imaginary parts of $\bar{\mu}$ and $\overline{K}_b$ on the basis of the experimental data
in this figure. Since we do not yet have a complete set of direct
experimental values for the log decrement corresponding to drained,
bulk motion, we have assumed that the log decrement for extensional
motion, $\delta_E = \pi E_i / E_r$ (where $E^* = E_r + iE_i$ is the drained Young's modulus)
undergoes a variation with frequency similar to that of δ_G and
falls somewhere in the range

$$0.9 \leq \delta_E / \delta_G \leq 1.5 \tag{2.2}$$

Eq. 2.2, which will be discussed more fully in Chap. 4, is based
on experimental and theoretical work described in Stoll (1985).
When the variation of δ_G with frequency from Fig. 2.7 and the
spread of δ_E / δ_G from Eq. 2.2 are used to pick a range of values
to be introduced into the frequency equation (Eq. 1.24) we obtain
the speckled band shown in Fig. 2.8. Thus we anticipate small
but significant changes in the logarithmic decrement for p-waves
to occur at much lower frequencies than predicted by the basic
Biot theory.

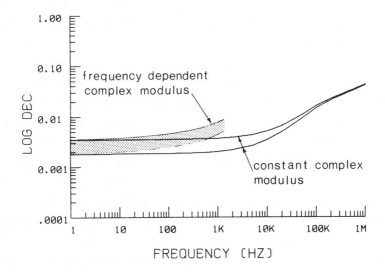

Fig. 2.8. Logarithmic decrement for p-waves in silt (Stoll, 1986).

REFLECTION AT SEDIMENT INTERFACES

In addition to the effects on velocity and attenuation illustrated by Figs. 2.1 through 2.7, the Biot model predicts many important differences in the acoustic field near boundaries between water and sediment or between sediments of different impedance. For example, the reflection coefficients at a plane interface between two sediments become frequency dependent and each incident wave is partitioned into three refracted and three reflected waves. However, in a typical water saturated sediment, the amount of energy converted to p-waves of the second kind is very small (Stoll, 1977,1980). Thus the bulk of the energy is refracted and reflected as p-waves of the first kind or shear waves.

Plane-wave reflection and refraction coefficients are generally based on the partitioning of energy at the interface between water and an elastic solid or between two elastic solids. The result is the classical Rayleigh reflection coefficient and other equations that relate the amplitudes of homogeneous incident waves and homogeneous reflected and refracted waves. For the case of an elastic solid, only two kinds of body wave can propagate and particle motion is either parallel or perpendicular to the wave normal depending on whether a dilatational or shear wave is being considered. The angles of incidence and emergence are related by Snell's law and, when one of the angles corresponding to a reflected or refracted wave increases to 90°, a "critical" angle of incidence is defined for the generating wave. When the angle of incidence of the generating wave exceeds a critical angle, the body wave which has become parallel to the interface no longer propagates and an interface wave is necessary to satisfy the boundary conditions. This wave decays exponentially away from the interface and its phase velocity is determined by the phase velocity of the generating wave projected onto the interface.

If we replace the elastic material with a viscoelastic or porous viscoelastic material, there are basic differences in the response at an interface. Reflected and refracted waves are generally inhomogeneous in the sense that the wave amplitude varies in the plane of constant phase and the trajectory of particle motion is elliptic in shape rather than parallel or perpendicular to the direction of the wave normal. A considerable body of literature has been published in recent years on the propagation of plane viscoelastic waves (Cooper and Reiss, 1966, Cooper 1967, Schoenberg, 1971, Buchen, 1971, Bocherdt, 1973, 1977) and some work has been done on reflection and refraction for the case of the Biot model (Deresiewicz and Rice, 1962, Geertsma and Smit 1961, Stoll, 1977, Stoll and Kan, 1981). Appendix B contains a derivation of the equations and boundary conditions that are necessary to study reflection and refraction of waves in a porous viscoelastic sediment. In the balance of this chapter, we present some results that illustrate the differences between the response of elastic materials and materials modeled by the Biot theory as implemented for marine sediments.

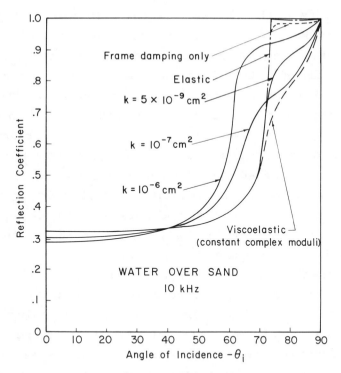

Fig. 2.9. Reflection coefficients for water over sand at 10 kHz. (Stoll and Kan, 1981).

We consider first plane, homogeneous waves incident in water to a layer of coarse sediment such as sand. In Fig. 2.9, the calculations are for a material with porosity 0.47 and frequency equal to 10 kHz. This figure shows the reflection coefficient at different angles of incidence for several different values of the coefficient of permeability along with a curve for an "elastic" sediment. The reflection coefficient is the ratio of the amplitude of the potential for reflected waves divided by the amplitude of the potential for the incident wave, A_r/A_i, and each curve is labeled with the appropriate coefficient of permeability. The elastic case is based on the same material properties except that the permeability is assumed to approach zero so that ζ, the increment of fluid content, also approaches zero. In addition damping in the skeletal frame is suppressed so that H and μ are the same as the real parts of $\overline{H}$ and $\overline{\mu}$ in the complete model. In

the example shown, the velocity of the dilatational wave of the
first kind is somewhat larger than the water velocity, so that
for the elastic case there is a critical angle of 73°.

Fig. 2.9 clearly shows that there is a significant difference
between the elastic response and the response of the "lossy"
sediment, which does not exhibit a "critical angle" in the usual
sense for any of the cases shown. Since overall attenuation in
the sediment model depends on both damping in the skeletal frame
and the frequency-dependent viscous losses in the fluid, it is
instructive to examine the relative contribution of these two
sources of dissipation. For this purpose we allow the permeability
to approach zero but retain the imaginary parts of the complex
moduli $\bar{H}$ and $\bar{\mu}$. Thus the viscous losses are eliminated and the
overall attenuation reflects only the influence of energy loss
in the skeletal frame. The result is the dashed line labeled
"frame damping only". This curve is essentially coincident with
the elastic case until after the critical angle is exceeded where
a small reduction of reflectivity is apparent. By neglecting the
viscous losses in the model, the overall attenuation expressed
as a logarithmic decrement is about .02 at all frequencies. This
is much too low to be realistic except at very low frequencies
so that the importance of the viscous losses in the coarse-grained
soils is obvious.

If the permeability of the sediment is decreased, there is
a progressive shift of the reflection curve toward the curve
labeled "frame damping only". As the permeabilities were changed,
the pore size parameter, a, was also changed in proportion to the
square root of the permeability while all other parameters were
held constant. Finally, when the sediment is modeled as an
ordinary viscoelastic material, where the overall moduli are
chosen to be complex constants, we obtain the reflection coefficients as shown by the broken line labeled "viscoelastic" in
Fig. 2.9. In this case there is only one kind of dilatational
wave and the attenuation depends directly on the ratio of imaginary
to real parts of the overall complex modulus. This approach has
been used in previous work where similar reflection curves have
been generated (Brekhovskikh, 1960, Mackenzie, 1960). In the
present example the curve labeled "viscoelastic" corresponds to

a constant complex modulus with a real part the same as for the
elastic case but with the imaginary part chosen so that the log
decrement for the dilatational is about 0.1.   Lower values of
decrement shift the curve toward the curve labeled "frame damping
only", which corresponds to a logarithmic decrement of .02 as
mentioned above.

     By comparing the "viscoelastic" curves with the curves for
the complete model, where the fluid mobility is taken into account,
it is clear that there is a significant difference over a wide
range of incident angle.   The difference between the two models
becomes smaller as the fluid mobility decreases so that in many
sediments of low permeability the simpler viscoelastic model may
be a satisfactory approximation.

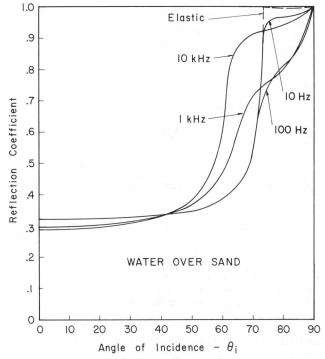

Fig. 2.10. Reflection coefficients for water over sand
at all angles of incidence for four different frequencies
(Stoll and Kan, 1981).

     Figs. 2.10 and 2.11 show the effect of frequency on the
reflection coefficient at a constant permeability of $10^{-6}cm^2$.
Below a frequency of about 100 Hz the properties of the skeletal

frame are important and the viscous losses somewhat less significant; however, as the frequency increases, the fluid losses
begin to dominate the overall attenuation and there are significant
differences from the elastic or purely viscoelastic response. The
reflection coefficients begin to deviate from the elastic case
at angles well below critical and the marked dependence on frequency
becomes apparent.   Thus at certain angles of incidence, the
reflection coefficient becomes strongly dependent on frequency
and an interface may act as a filter with respect to broadband
energy.

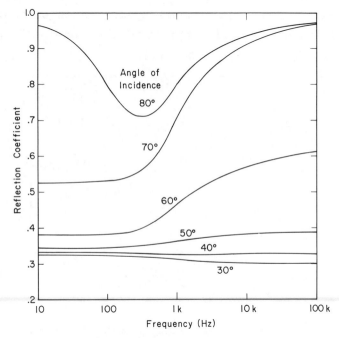

Fig. 2.11. Reflection coefficients for water over
sand at all frequencies for several different angles
of incidence (Stoll and Kan, 1981).

The reflection coefficients at normal incidence for a p-wave
in water over sand are shown in Fig. 2.12. This figure also shows
the effect of introducing a small amount of gas into the pore
water with the result that the effective bulk modulus of the water
is reduced by one order of magnitude (from 2×10^{10} to 2×10^{9}
dynes/cm^2). In the calculations for this latter case no correction
was made to account for the effect of the gas on fluid viscosity.
As might be expected, some dispersion in reflection coefficient

occurs in the same frequency range as the dispersion in velocity (see Fig.2.3, first and second curves).   The rather significant change in reflectivity caused by the presence of gas would obviously be important in cases where one is searching for trapped gas deposits or regions of biogenic gas production.

Finally we consider the case where a p-wave is propagating through a sequence of sediment layers with somewhat different impedance.   Fig. 2.13 shows the reflection coefficient for a normally incident dilatational wave in "mud" over sand.   The mud has a lower bulk modulus than the sand and a coefficient of permeability that is several orders of magnitude less than the permeability of the sand.   Because of the fact that two kinds of dilatational wave are generated at each interface, we are interested in the amount of energy that is converted to p-waves of the "second kind" and therefore presumably lost from the ongoing scattering processes because of the very high intrinsic damping in waves of the second kind.   The energy lost to this kind of wave, as a percentage of the total energy incident to the interface, is shown by dashed lines in Fig. 2.13.   As can be seen from the figure, a small but significant amount of energy is lost near each interface once a certain frequency level is reached.

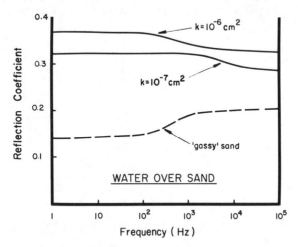

Fig. 2.12. Reflection coefficients for incident dilatational wave in water over sand - broken curve shows effect of a small amount of gas in the pore-water of the sand layer (Stoll, 1977)

For cases involving a number of layers or multiple reflections, the amount of energy lost from this kind of conversion could be considerable. Moreover, in situations where the presence of dissolved or free gas justifies modeling the fluid component as being more compressible than pure water, one expects to see a more significant fraction of the total energy converted to this kind of wave and then rapidly dissipated. There are many geological strata that contain significant amounts of gas in one form or another so the model will prove very useful in studying these cases.

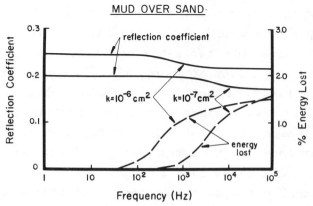

Fig. 2.13. Reflection coefficients and amount of energy converted to type 2 dilatational waves – mud over sand (Stoll, 1977).

CHAPTER 3
IDEALIZED GRANULAR MEDIA

INTRODUCTION

A good deal of insight into how a granular sediment responds to various kinds of loading can be obtained by reviewing the work that has been done on the mechanics of regular arrays of discrete particles. Starting in the early 1950's, Mindlin (1954), Duffy and Mindlin (1957), Dereziewicz (1958) and a number of other authors studied the response of various kinds of regular arrays of elastic spheres. Various packings such as the simple cubic, the face-centered cubic and the hexagonal close-packed array have been analyzed by considering the known geometry of the packing and the intergranular forces between particles. The distribution of tractions on the contact areas between particles was first studied by Hertz (see Timoshenko and Goodier, 1951), Cattaneo (1938) and Mindlin (1949) with the work of Hertz predating that of the other two authors by over 100 years. However, Hertz's work dealt only with normal forces between the particles, whereas the work of Mindlin and Cattaneo considered the effect of tangential tractions, setting the stage for the development of models to describe the complete response of different arrays of particles. We begin our discussion with a brief summary of contact mechanics and continue with a discussion of the equilibrium and compatibility requirements that must be enforced in order to model various arrays.

CONTACT MECHANICS

When two like spheres of elastic material are pressed together in such a manner that the contact forces are normal to the plane of contact, the stresses over the small area of contact and the normal compliance are given by the Hertz theory. For spheres of radius R, shear modulus μ, and Poisson's ratio ν, the radius of the circular area of contact, a, is given by

$$a = \left[\frac{3(1-\nu)RN}{8\mu} \right]^{1/3} \qquad (3.1)$$

where N is the magnitude of the normal contact force. If the relative approach of the sphere centers is denoted by α, then the normal compliance, C, is

$$C = \frac{d\alpha}{dN} = \frac{1-\nu}{2\mu a} \qquad (3.2)$$

and the distribution of normal traction over the area of contact is given by

$$\sigma = \frac{3N}{2\pi a^3}(a^2 - \rho^2)^{1/2} \qquad (3.3)$$

where ρ is the radial distance from the center of the contact circle. The distribution of traction given by Eq. 3.3 is shown in Fig. 3.1.

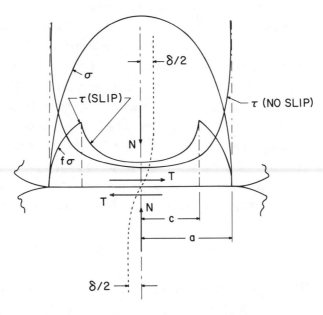

Fig. 3.1. Distribution of normal and tangential tractions on the contact surface of a pair of spherical particles (Mindlin, 1954)

If two spheres are pressed together by a constant normal force and then tangential forces are applied in the plane of contact, a linear elastic model predicts that the pattern of

normal traction, σ, remains unchanged while the resulting distribution of tangential tractions, τ, tends to infinity at the
periphery of the circular contact area (see Fig. 3.1). Since
infinite stresses are not physically possible, stress relief of
some sort must occur in the regions of stress concentration. To
allow this relief, we assume that relative displacement between
contiguous points within the contact area between two spheres
occurs when the Coulomb law of friction is satisfied

$$\tau = f\sigma \tag{3.4}$$

where f is the coefficient of sliding friction. As a result, an
annular ring of slip tends to form at the outer edge of the contact
area. The annulus of slip expands radially inward as the ratio
T/N increases until Eq. 3.4 is satisfied over the entire contact
area. At this point T = fN and relative displacement or "sliding"
occurs over the entire area of contact. For ratios of T/N, 0 <
T/N < f, the inner radius, c, of the annular zone of slip is

$$c = a\left(1 - \frac{T}{fN}\right)^{1/3} \tag{3.5}$$

and the tangential compliance is given by

$$S = \frac{d\delta}{dT} = \frac{2-\nu}{4\mu a}\left(1 - \frac{T}{fN}\right)^{-1/3} \tag{3.6}$$

where δ is the relative displacement of the sphere centers in the
direction parallel to the plane of contact. The distribution of
tangential traction both with and without slip is shown in Fig.
3.1.

When both T and N are changing simultaneously, the expressions
for tangential compliance are much more complicated, taking the
form of nonlinear differential equations. For example, assume
that after applying a normal force both N and T are increased at
an arbitrary relative rate. Then instead of Eq. 3.6 the tangential
compliance is given by

$$S = \frac{2-\nu}{4\mu a}\left[f\frac{dN}{dT} + \left(1 - f\frac{dN}{dT}\right)\left(1 - \frac{T}{fN}\right)^{-1/3}\right], \qquad 0 \le \frac{dN}{dT} \le 1/f$$

$$S = \frac{2-\nu}{4\mu a}, \qquad \frac{dN}{dT} \ge 1/f \tag{3.7}$$

The necessary rules for deriving the appropriate equations for
a variety of loading and unloading paths were established in an
exhaustive paper by Mindlin and Deresiewicz (1953). We will use
these rules subsequently to derive an expression for tangential
compliance during the loading and unloading of a regular array
of spheres.

REGULAR ARRAYS

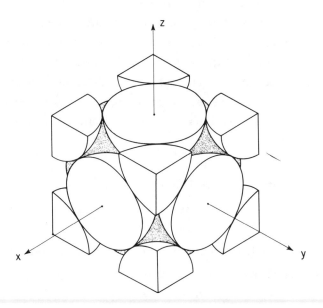

Fig. 3.2 Schematic showing the arrangement of spheres
in a face-centered cubic (Stoll, 1989)

With the aid of the equations for compliance discussed above,
a number of different investigators have studied regular arrays
of spheres in an effort to determine stress-strain relationships,
wave velocities and failure criteria. Various arrays that have
been studied include the simple cubic, the hexagonal close-packed,
and the face-centered cubic. In this chapter we concentrate on
various kinds of axisymmetric loading on a face-centered cubic
(fcc) array and rely heavily on the prior work of Duffy and Mindlin
(1957) and Hendron (1963).

The fcc is one of the densest regular arrays, with a coordination number of 12 (i.e., each sphere is in contact with 12 neighboring spheres) and a porosity of 25.95%. Fig.3.2 is a schematic diagram of the smallest repeating element of volume in a fcc and Fig. 3.3 shows the components of the contact force at each of the 12 points of contact with neighboring spheres.   Of the 36 components shown on Fig.3.3, only 18 are independent in the general case and because of symmetry, only three are required for the case of axisymmetric loading.

$$T_{xx} = T_{xx}' = T_{yy} = T_{yy}' = T_{zz} = T_{zz}' = 0$$

$$T_{xy} = T_{xy}' = 0$$

$$N_{xy} = N_{xy}' = N_1$$

$$N_{zy} = N_{yz}' = N_{zx} = N_{zx}' = N_2 \qquad\qquad (3.8)$$

$$T_{zx} = T_{zx}' = -T_{yz} = -T_{yz}' = T_2$$

For the general case, one can write 9 equations of equilibrium in terms of the 18 independent components of contact force and 9 compatibility equations in terms of the relative displacements of the centers of the spheres.   Again, because of the many symmetries, only three of the components of relative displacement are needed. As a result, only two equations of equilibrium

$$dN_2 + dT_2 = \frac{\sqrt{2}}{8} dP_{zz} = \sqrt{2} R^2 d\sigma_{zz} \qquad\qquad (3.9)$$

$$dN_1 + dN_2 - dT_2 = \frac{\sqrt{2}}{4} dP_H = 2\sqrt{2} R^2 d\sigma_h$$

and one compatibility equation

$$d\alpha_1 - d\alpha_2 + d\delta_2 = 0 \qquad\qquad (3.10)$$

are required to define the problem of axisymmetric loading.   In the above equations P_{zz} and P_h are the total normal forces on the horizontal and vertical faces of the cube shown in Fig. 3.2 and the relative displacements in Eq. 3.10 are related to the force components in Eq. 3.9 by the compliances

$$d\alpha_1 = C_1 dN_1$$

$$d\alpha_2 = C_2 dN_2 \qquad\qquad (3.11)$$

$$d\delta_2 = S_2 dT_2$$

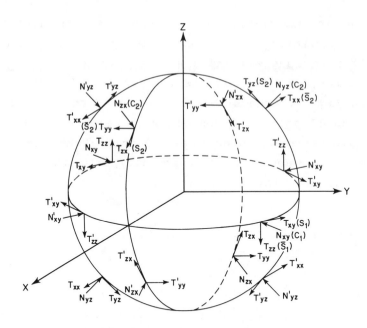

Fig. 3.3. Contact forces acting on a typical sphere
of a face-centered cubic (modified from Thurston and
Deresiewicz, 1959)

Equations 3.9, 3.10, and 3.11 are sufficient to describe the
response of the fcc array under axisymmetric loading provided the
vertical and lateral forces P_{zz} and P_h are specified. However,
for the case of uniaxial strain, it is necessary to specify that
the radial strains are zero and solve for the lateral forces.
When all of the symmetries of the problem are taken into account,
the expressions for increments of lateral and vertical stress and
strain take the form

$$d\epsilon_h = \frac{d\alpha_1}{2R} = \frac{C_1 dN_1}{2R}$$ (3.12)

$$d\epsilon_{zz} = \frac{1}{2R}(d\alpha_2 + d\delta_2)$$

Thus, Eqs. 3.9, 3.10, 3.11, and 3.12 are sufficient to define the
problem of a fcc array loaded by a vertical force P_{zz} and restricted
to symmetrical lateral deformations. In the case of uniaxial
strain in the z direction, $d\epsilon_h = 0$ for all loading histories.

Hendron (1963) used the above equations together with Eqs. 3.2 and 3.7 to describe the stress-strain relationship for a fcc array subjected first to an isotropic stress of arbitrary level followed by one-dimensional strain. By a limiting process, he deduced from his solution that during initial monotonic loading from zero stress, one-dimensional compression would cause sliding at all contacts corresponding to normal and tangential forces N_2 and T_2, and therefore that T_2 is related to N_2 by the relationship

$$T_2 = f N_2 \qquad\qquad (3.13)$$

over the entire initial monotonic loading path. On the basis of this important result, he derived the following stress-strain relationship for the initial loading

$$\epsilon_{zz} = \left[\frac{1-\nu}{\mu(1+f)} \frac{3}{2} \right]^{2/3} \sigma_{zz}^{2/3} \qquad\qquad (3.14)$$

The loading history used to obtain Eq. 3.14 (uniaxial strain in the z direction) corresponds to the stress path that occurs during normal deposition of sediments wherein the stress is the result of a continually increasing overburden pressure; in geotechnical vernacular, the resulting stress-strain curve is known as the "virgin consolidation curve". One is tempted to take the derivative of Eq. 3.14 in order to obtain a modulus for computing the wave velocity in the z direction. However, it is well known that the slope of the "virgin consolidation curve" does not yield the correct modulus for computing the velocity of dilatational waves of small amplitude (Stoll, 1968). In fact, it has been shown in many experiments that at any stress level arrived at by monotonic loading, it is the slope of an unloading path at the point where unloading begins that matches the dynamic modulus of interest. This is because there is a strong discontinuity in the slope of the stress-strain curve at the point where unloading is initiated. Thus small oscillations are controlled by the slope of the unloading curve at any point after a monotonic loading. Moreover, if we now perform monotonic unloading to some point, the dynamic modulus for small oscillations will depend on the initial slope of a reloading curve initiated at that point. For this reason, we will derive expressions for the unloading-reloading paths that will result if the direction of loading is reversed at any arbitrary point.

Before looking at unloading and reloading paths, it is
instructive to compare the stress-strain relationship for uniaxial
strain given by Eq. 3.14 with the equation that one obtains for
the case of isotropic loading. When $\sigma_{zz} = \sigma_h = \sigma_0$, $N_1 = N_2$ and $T_2 = 0$
so that there are no tangential forces on any of the grain contacts
and so the problem is statically determinate. Under these conditions the relationship between total stress and strain is given
by (Hendron, 1963)

$$\epsilon_{zz} = \left[\frac{3(1-\nu)\sqrt{2}}{8\mu} \right]^{2/3} \sigma_0^{2/3} \tag{3.15}$$

As might be expected, the sphere pack is stiffer during isotropic
loading than during one-dimensional compression.

After monotonic, isotropic loading to any stress level, the
incremental stress-strain relationship for the fcc exhibits cubic
symmetry and the wave velocities in various directions are easily
calculated (e.g., see Stoll, 1985 and Duffy and Mindlin, 1957).

UNLOADING AND RELOADING AFTER MONOTONIC LOADING

For the case of uniaxial strain, Hendron's work showed that
at any point on the initial loading path given by Eq. 3.14, the
normal and tangential tractions at the contact points corresponding
to the forces N_2 and T_2 are always given by the relation $\tau = f\sigma$
everywhere on the contact surface so that the actual distribution
of τ may be determined with the aid of Eq. 3.3. In order to solve
Eqs. 3.9 through 3.13 starting with this distribution of τ, we
make use of the rules given by Mindlin and Deresiewicz (1953) for
determining the tangential compliance at a point of contact between
spheres for the case where both N and T are decreasing at arbitrary
rates. If N^* and T^* are the values of N and T at the point where
unloading is begun, these rules show that the compliance must be

$$\frac{d\delta}{dT} = \frac{2-\nu}{4\mu a} \left[-f\frac{dN}{dT} + \left(1 + f\frac{dN}{dT} \right) \left(1 - \frac{T^*-T}{2fN} + \frac{N^*-N}{2N} \right)^{-1/3} \right] \tag{3.16}$$

Then, still assuming that $\epsilon_h = 0$ (uniaxial strain), Eqs. 3.2, 3.9 through 3.13, and 3.16 are used to obtain the following differential equation that determines the relationship between N_2 and T_2 during the unloading

$$\frac{dT}{dN} = -f + \frac{1 + k_1 f}{k_1} \left(1 - \frac{T^* - T}{2fN} + \frac{N^* - N}{2N} \right)^{1/3} \tag{3.17}$$

where

$$k_1 = \frac{2 - v}{2(1 - v)}$$

Integration of Eq. 3.17 gives

$$T - T^* = 2f \left[\left((1 + k_2)(N^*)^{2/3} - k_2 N^{2/3} \right)^{3/2} - \frac{N^* + N}{2} \right] \tag{3.18}$$

and finally eliminating N and T between the equations for stress and strain increments (Eqs. 3.12), we derive the following stress-strain relationship for the unloading path.

$$\sigma_{zz} = \frac{\mu}{1 - v} \frac{2}{3} \left\{ 2f \left[(1 + k_2)\epsilon^*_{zz} - k_2\epsilon_{zz} \right]^{3/2} + (1 - f)(\epsilon_{zz})^{3/2} \right\},$$

$$[(1 + k_2)\epsilon^*_{zz} - k_2\epsilon_{zz}] \geq 0 \tag{3.19}$$

$$\sigma_{zz} = \frac{\mu}{1 - v} \frac{2}{3}(1 - f)(\epsilon_{zz})^{3/2}, \qquad [(1 + k_2)\epsilon^*_{zz} - k_2\epsilon_{zz}] < 0$$

where ϵ^*_{zz} is the strain at the point where unloading is initiated and $k_2 = -(1 + k_1 f)/k_1/2f$.

Fig. 3.4 shows the stress-strain curves produced by Eqs. 3.12 and 3.19 for unloading paths starting at two different stress levels after monotonic loading. Before proceeding with a study of wave velocities associated with the unloading path, it is instructive to examine the distribution of tangential tractions that occur at the grain contacts corresponding to the contact forces N_2 and T_2 during unloading. Fig. 3.5 shows the distributions of tangential traction that correspond to several points along the unloading path labeled A-B-C-D-E in Fig. 3.4. At the beginning of unloading , point A, the tangential traction is proportional to the normal traction at all points on the intergranular contact surface. As the unloading proceeds, the radius of the contact area decreases, slip begins to occur at the outer edge of the contact area (with direction of relative motion opposite to that

occurring during loading) and the distribution of traction changes
as shown by the succession of curves labeled B,C,D, and E.   At
point E, the annulus of slip has enlarged to cover the entire
contact surface and sliding in a direction opposite to that
occurring during the initial monotonic loading is about to begin.
This corresponds to the condition $(1+k_2)\epsilon'_{zz} = k_2\epsilon_{zz}$ in Eq. 3.19.
Thus, during the remainder of the unloading path, from E to the
origin, the distribution of tangential traction is determined by
Eq. 3.13, with the direction of the tangential force, T, reversed
from the direction during the virgin loading.

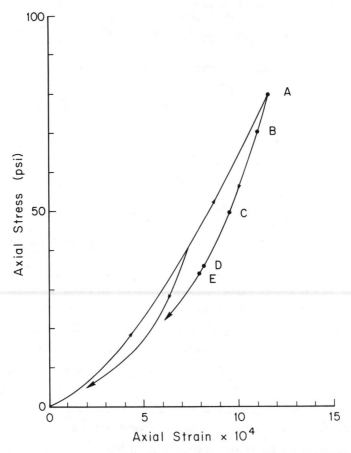

Fig. 3.4.   Stress-strain curve for uniaxial strain
in the direction of a principal axis in a fcc array
(Stoll, 1989).

The one-dimensional compression test (often called an oedometer
or consolidation test) has been used for many years by geotechnical

engineers to study the compressibility of soils. An example of
the results of such a test, performed on a sample of 20-30 Ottawa
sand, is shown in Fig. 3.6 (Stoll, 1989) The apparatus used in
this test was specially designed to permit the measurement of
radial stresses and to prevent the deleterious effects of side
friction that often affect consolidation tests utilizing stiff
rings for lateral confinement.   The apparatus consisted of a
cylindrical rubber shell in which a very thin band of spring brass
had been molded so as to be in direct contact with the sediment.
Strain gauges bonded to this metal band sense any tendency for
radial strain to occur and the radial confining stress, produced
either by an internal vacuum or an external pressure, is adjusted
to hold the diameter of the specimen constant.

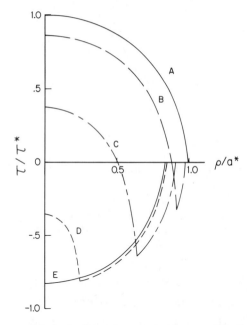

Fig. 3.5. Distribution of tangential traction on
certain contact areas between spheres during uniaxial
strain. The curves labeled A through E correspond to
the points on the stress-strain curve in Fig. 3.4.
(Stoll, 1989)

Ottawa sand is a naturally occuring sand with rounded, uniform
grains and our test was performed on a sample with a porosity of
about 33.5%. This is quite a bit less than the porosity of a fcc
array (25.95%). Even more importantly, the inclinations of the

contact areas with respect to the principal loading direction
vary infinitely from 0 to 90° while in the fcc they are at either
0, 45 or 90°.   Thus, even though the general shape of the curve
is in agreement with the predictions of the model, the strain
amplitudes are markedly different as a result of much more
interparticle sliding in the random array of the Ottawa sand.

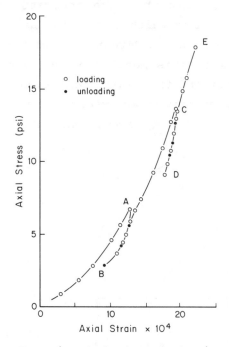

Fig. 3.6.   Experimental stress-strain curve for a
uniaxial compression test on 20-30 Ottawa sand - loading
sequence A,B,C,D,E (Stoll, 1989).

We next consider the effect of the initiation of reloading
at any point on the unloading path described above.   Again, based
on the work of Mindlin and Deresiewicz (1953), we find the
tangential compliance for a reloading path to be

$$\frac{d\delta}{dT} = \frac{2-\nu}{4\mu a}\left[f\frac{dN}{dT} + \left(1 - f\frac{dN}{dT}\right)\left(1 + \frac{T_L - T}{2fN} + \frac{N_L - N}{2N}\right)^{-1/3}\right] \qquad (3.20)$$

where N_L and T_L are the normal and tangential forces at the point
where reloading is initiated.   Thus, the tangential compliance
at the point of reloading where $N = N_L$ and $T = T_L$ is of the same
form as the compliance at the point where unloading was started

but with a different value of the contact radius "a" which can
be determined only by knowing the details of the unloading path
as presented above.

WAVE VELOCITIES

   We now have enough information to determine the compliance
for small oscillations of the load over the entire monotonic
unloading path so that it is possible to describe the propagation
of small amplitude waves through the prestressed matrix of spheres.
Duffy and Mindlin (1957) showed that after any axisymmetric
loading, the incremental stress-strain relationships for a fcc
array are of the form

$$d\sigma_{xx} = c_{11}d\epsilon_{xx} + c_{12}d\epsilon_{yy} + c_{13}d\epsilon_{zz}$$

$$d\sigma_{yy} = c_{12}d\epsilon_{xx} + c_{11}d\epsilon_{yy} + c_{13}d\epsilon_{zz}$$

$$d\sigma_{zz} = c_{13}d\epsilon_{xx} + c_{13}d\epsilon_{yy} + c_{33}d\epsilon_{zz} \qquad (3.21)$$

$$d\sigma_{yz} = 2c_{44}d\epsilon_{yz}$$

$$d\sigma_{zx} = 2c_{44}d\epsilon_{zx}$$

$$d\sigma_{xy} = 2c_{66}d\epsilon_{xy}$$

where the compliances of the array, c_{ij}, are given in terms of the
compliances of the individual contacts by

$$c_{11} = (C_1^{-1} + S_1^{-1} + C_2^{-1} + S_2^{-1})/2\sqrt{2}R$$

$$c_{12} = (C_1^{-1} - S_1^{-1})/2\sqrt{2}R$$

$$c_{13} = (C_2^{-1} - S_2^{-1})/2\sqrt{2}R$$

$$c_{33} = (C_2^{-1} + S_2^{-1})/\sqrt{2}R \qquad (3.22)$$

$$c_{66} = \left(C_1^{-1} + \overline{S}_2^{-1}\right)/2\sqrt{2}R$$

$$c_{44} = \frac{1}{2\sqrt{R}}\left[\frac{1}{C_2} + \frac{2S_2 + \overline{S}_1 + \overline{S}_2}{2\overline{S}_1\overline{S}_2 + S_2(\overline{S}_1 + \overline{S}_2)}\right]$$

For the loading path which we are considering (uniaxial strain
in the z direction) the compliances for the individual contacts
are

$$C_1 = S_1 = \overline{S}_1 = \infty$$

$$C_2 = C \tag{3.23}$$

$$S_2 = \overline{S}_2 = S$$

In Eq. 3.22, C can be obtained from Eq. 3.2 while S will be of the form

$$S = \frac{2-\nu}{4\mu a} \tag{3.24}$$

obtained by letting $T = T_*$ and $N = N_*$ in Eq. 3.16 or $T = T_L$ and $N = N_L$ in Eq. 3.20. Again, the value of a depends on the entire loading history.

If we now consider plane body waves of small amplitude moving through the prestressed fcc array in either the vertical or horizontal directions, we may calculate the wave velocity from the following expressions which are the same as those applicable to a crystal with tetragonal symmetry.

Compressional waves:

$$c_z = \sqrt{c_{33}/\rho}$$

$$c_x = c_y = \sqrt{c_{11}/\rho}$$

$$c_z/c_x = 1.41$$

Shear waves:

$$s_{yz} = s_{zx} = \sqrt{c_{44}/\rho}$$

$$s_{xy} = \sqrt{c_{66}/\rho}$$

$$s_{yz}/s_{xy} = 1.2$$

EFFECTS OF WATER SATURATION

In order to investigate the effects of water saturation, we utilize the Biot theory for porous materials implemented for the case where the moduli of the skeletal frame are considered to be transverse isotropic. For the transverse isotropic case, Biot defines the following stress-strain relations,

$$d\sigma_{xx} = B_1 d\epsilon_{xx} + B_2 d\epsilon_{yy} + B_3 d\epsilon_{zz} + B_7 d\zeta$$

$$d\sigma_{yy} = B_2 d\epsilon_{xx} + B_1 d\epsilon_{yy} + B_3 d\epsilon_{zz} + B_7 d\zeta$$

$$d\sigma_{zz} = B_3 d\epsilon_{xx} + B_3 d\epsilon_{yy} + B_4 d\epsilon_{zz} + B_8 d\zeta$$

$$dp_f = B_7 (d\epsilon_{xx} + d\epsilon_{yy}) + B_8 d\epsilon_{zz} + B_9 d\zeta \qquad (3.25)$$

$$d\sigma_{yz} = 2 B_5 d\epsilon_{yz}$$

$$d\sigma_{zx} = 2 B_5 d\epsilon_{zx}$$

$$d\sigma_{xy} = 2 B_6 \epsilon_{xy}$$

where the $d\sigma_{ij}$ are increments of total stress on an element of volume attached to the fcc array, the $d\epsilon_{ij}$ are increments of strain of the element of volume, dp_f is an increment of fluid pressure and $d\zeta$ is the incremental change in the increment of fluid content.

If we now rewrite Eqs. 3.25 in terms of intergranular stresses $\sigma_{ij}' = \sigma_{ij} + p_f$ and consider an isotropic loading of the array in a "drained, jacketed" test, we obtain the following rela- tionships between the stiffness constants of Eqs. 3.21, c_{ij}, and those of Eq. 3.25, B_k, assuming $p_f = 0$.

$$c_{11} = B_1 - \frac{B_7^2}{B_9}$$

$$c_{12} = B_2 - \frac{B_7^2}{B_9}$$

$$c_{13} = B_3 - \frac{B_7 B_8}{B_9} \qquad (3.26)$$

$$c_{33} = B_4 - \frac{B_8^2}{B_9}$$

$$c_{44} = B_5, \qquad c_{66} = B_6$$

The parameters B_7, B_8, and B_9 are strongly dependent on the compressibility of the fluid saturant and the solid material of which the particles are composed. Thus, in order to evaluate these constants, we need to consider a test where the fluid pressure is changing as in the "unjacketed" test. As described in Chap. 1, we assume that a saturated, unjacketed array of particles is surrounded by a fluid which may be pressurized to any desired level. Thus the total stresses in the array would be $\sigma_{xx} = \sigma_{yy} = \sigma_{zz} = -p_f = \sigma$, where σ is the fluid stress within the pore space of the array and p_f is the externally applied fluid pressure.

For this type of loading, the actual force transmitted particle to particle and therefore the resultant "intergranular" or "effective" stresses will depend on the sum of the projected areas of the contact "points" between particles per unit of total cross-sectional area of an element. In the case of an array with no prestress, the contact area is theoretically zero so that the unjacketed loading will cause no intergranular stresses, the volume strain is isotropic and, consequently, the porosity remains unchanged. Under these conditions the unjacketed compressibility of the array, δ, is the same as the compressibility of the solid particles so that Eq. 1.12 is applicable and the coefficient of fluid content $\gamma = \zeta/p_f$ is given by Eq. 1.11.

If, on the other hand, the array of particles is prestressed before the unjacketed test is performed , neither Eq. 1.11 nor Eq. 1.12 is strictly valid since some effective stress will be generated owing to the finite areas of intergranular contact. To see the effect of interparticle flattening, we may write the effective stress-strain relations in the form

$$\sigma_{xx}' = \sigma_{xx} + (1 - \alpha_i)p_f = c_{11}\epsilon_{xx} + c_{12}\epsilon_{yy} + c_{13}\epsilon_{zz} \tag{3.27}$$

where α_i is the ratio of the sum of the projected areas of contact in the x direction over the total area of the face of an element of volume attached to the array. In the case of anisotropic prestress, α_i will be different in different coordinate directions. However, for typical inorganic sediments, where individual grains may be assumed to deform isotropically under fluid pressure, (i.e., quartz or silt) and small overburden pressure (i.e., the equivalent of say 10 m of embedment under water), α_i is of the order of 10^{-4}. Thus in the unjacketed test where $\sigma_{xx} = \sigma_{yy} = \sigma_{zz} = -p_f$, the left-hand side of Eq. 3.27 will approach zero and $\epsilon_{xx} \doteq \epsilon_{yy} \doteq \epsilon_{zz}$.

In view of the above, we assume as a first approximation that $\epsilon_{xx} = \epsilon_{yy} = \epsilon_{zz}$ in unjacketed loading and use Eqs. 1.11 amd 1.12 to introduce the effects of fluid and solid compressibility and derive the following relations using the first four of Eqs. 3.26

$$B_7 = - \frac{1 - (c_{11} + c_{12} + c_{13})\delta/3}{\delta - (2c_{11} + 2c_{12} + 4c_{13} + c_{33})\delta^2/9 + \gamma}$$

$$B_8 = - \frac{1 - (2c_{13} + c_{33})\delta/3}{\delta - (2c_{11} + 2c_{12} + 4c_{13} + c_{33})\delta^2/9 + \gamma} \qquad (3.28)$$

$$B_9 = \frac{1}{\delta - (2c_{11} + 2c_{12} + 4c_{13} + c_{33})\delta^2/9 + \gamma}$$

For quartz spheres and water as a saturant, typical values for δ and γ are 2.78×10^{-11} (Pa^{-1}) and 4.72×10^{-10} (Pa^{-1}), respectively, whereas $(c_{11} + c_{12} + c_{13})$ and $(2c_{13} + c_{33})$ have values of 5.36×10^8 (Pa) and 10.73×10^8 (Pa) respectively for an overburden pressure of say 10^5 Pa (the equivalent of about 10 m of overburden under water). Using these values in Eqs. 3.25, 3.26 and 3.28 we may calculate the ratio of p-wave velocities in the x and z directions assuming closed system conditions (i.e., $\zeta = 0$)

$$c_z/c_x = \sqrt{\frac{B_4/\rho_T}{B_1/\rho_T}} = 1.002$$

where ρ_T is the total mass density. The ratio of the shear wave velocities is unaffected by the presence of the fluid.

From the above result, it is clear that water saturation of the fcc array tends to reduce the stress-induced anisotropy to very small values when the intergranular stress from the overburden is small. Higher levels of confining stress will increase the values of the c_{ij} and therefore the ratio of c_z to c_x. However, it should be remembered that the simplifying assumptions leading to our results above will be less valid since the areas of contact will increase as the confining stress increases. A discussion of the effects of prestress is given in Biot and Willis (1957).

SUMMARY AND SOME CONCLUSIONS

By taking advantage of the known geometry and many symmetries of a face- centered cubic array of spheres, we have shown that it is possible to derive the total stress-strain relationship that is valid over an entire loading-unloading path for the case of uniaxial strain. As a result, all components of the compliance tensor relating increments of stress and strain are known for points on the unloading path, and therefore it is possible to calculate the velocity of p- and s-waves propagating in different directions through the array. The prestress produced by the initial monotonic loading causes the incremental response to be anisotropic at any point on the unloading path and, as a result, p-wave velocities in the direction of initial axial strain (in the present case in the z direction) are found to be about 40% larger than in the transverse directions. In addition, horizontally polarized s-waves are about 20% slower than vertically polarized shear waves.

When the effects of fluid saturation are incorporated into the model with the help of the Biot theory, the ratio of vertical to transverse p-wave velocities is found to decrease to a value nearly equal to 1. Thus the relatively low compressibility of the fluid relative to the compressibility of the skeletal frame tends to greatly reduce the degree of anisotropy produced by the initial quasistatic loading.

Since the entire response of our simple model is governed largely by the compliance of a single set of contact points, with planes of osculation inclined at an angle of 45° to the horizontal, it is clear that the calculated results are, at best, a rough approximation to the response of real sediments composed of random arrays of particles of varying size and shape. However the effects of basic interparticle mechanics that govern the response of all particulate materials are clearly evident in the nonlinear, path-dependent response. Moreover, the predicted degree of anisotropy is in general agreement with the limited experimental results that are currently available.

One of the most important insights that we can gain from the discussion in this chapter is that the dynamic modulus (stiffness under small, oscillatory loading) of any particulate material depends largely on the number of interparticle contacts per unit of volume and the area of each contact. Since the area of contact varies as the 1/3 power of the normal contact force (Eq. 3.1), all of the models based on regular arrays as well as those made up of random arrays (e.g., Brandt, 1955) predict a dynamic shear modulus proportional to the effective stress to the 1/3 power (e.g., differentiate Eq. 3.18, let $\epsilon_{zz}^* = \epsilon_{zz}$ and solve for $d\sigma_{xx}/d\epsilon_{zz}$).

Experimental results generally indicate that the modulus increases
with decreasing porosity and depends on the confining pressure
to a power somewhere in the range of 1/3 to 1/2, with the value
closer to 1/3 as the stress level increases.   Thus the very
important effect of porosity and overburden pressure,   which
determines the number of intergranular contacts per unit volume
and the area of each, is implied by our simple model studies.

The quantitative difference between theory and experiment
is due largely to the fact that in random arrays of real sediments
(or for that matter, spheres) increasing the confining stress
increases the number of contacts per sphere.   This same effect
is also seen in experiments with "regular" arrays of like spheres
since the manufacturing tolerances always result in some small
variations in diameter.   Thus a fcc array may not truly have a
coordination number of 12 until high confining stress is applied.
This effect was shown experimentally by Duffy and Mindlin (1957).

CHAPTER 4
LABORATORY EXPERIMENTS

INTRODUCTION

In this chapter we describe some experiments which were carried out to test certain predictions of the mathematical model and to provide data helpful in estimating some of the required parameters. To accomplish these goals several experimental configurations with new and unique features were designed to allow careful control of the variables that previous research has shown to be important in wave propagation through sediments. Of these variables the three that are generally considered to be the most important, at least in coarser sediments, are the mean effective stress, the voids ratio, and the amplitude of cyclic shear strain (Hardin and Drnevich, 1972). In addition, we wished to study the effects of frequency over the widest possible range without tampering with the specimen or boundary conditions in our experiment. The ability to observe the response over a wide range of frequencies is important because the Biot theory, as implemented for marine sediments, predicts a significant effect of frequency that has not been studied in previous experiments or accounted for in other theories.

EARLY RESONANT COLUMN EXPERIMENTS

A schematic diagram of the experimental setup used in our first series of tests (Stoll, 1979) is shown in Fig. 4.1 along with an example of the kind of output that is recorded. The specimen of sediment used in these experiments was confined in a thin, cylindrical shell composed of a spiral of steel wire embedded in latex rubber. The wire was 0.25 mm in diameter and the spacing between adjacent loops was also approximately 0.25 mm. The inside diameter of the shell was 4.9 cm and the length of each specimen was 27 cm. The shells were designed to offer very high resistance to radial deformation, but virtually no

resistance to twisting or to compression or extension in the
axial direction. The total weight of each shell was about 60 g
while the total dry weight of a typical specimen averaged 900 g.

The bottom of the reinforced shell is sealed to a cylindrical
pedestal fixed to a rigid base. A porous plate, made of sintered
bronze particles, is attached to the top of the pedestal to allow
drainage and measurement of pore-water pressure. In addition,
there is a tiny hole through the pedestal containing an "O" ring
seal that allows a thin stainless steel wire, running the length
of the specimen, to pass through the pedestal without leakage of
pore fluid and with virtually no friction on the wire. The axial
wire, which is attached to the center of the top cap of the
specimen, is used to apply a static axial stress to the specimen.
Different axial stress levels are attained by placing weights on
a platform attached to the lower end of the wire beneath the rigid
base.

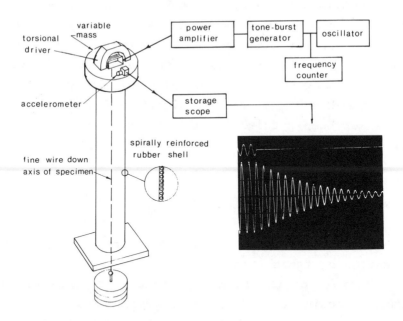

Fig. 4.1. Schematic of experimental setup for studies of
torsional vibration (Stoll, 1979).

The upper mass, which moves with the top of the specimen,
is composed of a cap with a rough bottom (sand particles embedded
in epoxy cement) in direct contact with the sediment and a circular

plate which supports several transducers and a lightweight driver coil. In addition, one or more annular rings made of brass may be attached to the plate in order to vary the mass moment of inertia of the top assemblage. Thus the specimen and top mass form a torsional pendulum with a resonant frequency that depends on the stiffness of the specimen and the mass moment of inertia of the top.

The use of an articulated shell and a mechanical loading system, in lieu of the more commonly employed resonant column apparatus with a pressure chamber and rubber membrane, was adopted because it allowed much easier preparation of fully saturated specimens and stable static loading over long periods of time. The advantages of this approach were felt to outweigh some of the disadvantages which include small perturbing effects of the shell in the form of added mass as well as some uncertainty in determining the radial stress for comparison with other experimental results.

As can be seen from Fig. 2.1, our theoretical model predicts that the logarithmic decrement for shearing motion will depend rather strongly on the coefficient of permeability, and the maximum damping caused by relative fluid motion is shifted to higher and higher frequencies as the permeability decreases. For relatively coarse sands, with permeabilities of the order of 10^{-6} cm^2, the effects of viscous losses begin to appear in the range of 1 to 10 Hz, whereas for finer materials such as silt or clay, these effects do not show up until much higher frequencies are reached (i.e., for silt with a permeability of the order of 10^{-9} cm^2, the effect of viscous losses is insignificant until frequencies are over 10 kHz). For this reason several different gradations of sediment were tested in order to insure a sufficient spread of permeability to test the predictions of the model.

Grain size curves for four different sediments used in the tests are shown in Fig. 4.2. Sample No. 1 is "20-30" Ottawa sand which is standard testing sand with rounded grains. This material was chosen for one series of tests because there is a great deal of data in the literature that can be compared with our results (e.g., see Hardin and Richart, 1963 and Hardin, 1965). Sample No. 2 is a coarse beach sand from Fire Island, New York. Both samples 1 and 2 have about the same range of grain size; however,

the Fire Island sand has a smaller mean diameter and the grains
are more angular. Sample No. 3 is a naturally occurring medium
to fine sand with a smaller mean diameter and a larger range of
grain sizes than either No. 1 or No. 2. Sample No. 4 is an
inorganic, micaceous silt (rock flour) obtained from a glacial
deposit in Long Island Sound, New York.

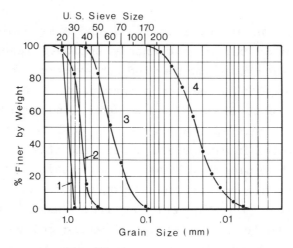

Fig. 4.2. Grain size curves for sediments used in
experiments (Stoll, 1979).

Several series of tests were performed on each different
kind of sediment, some with a dry specimen and others with the
specimen fully saturated with water. In order to insure 100%
saturation in the wet tests, the testing shell was first filled
with water after which wet sediment was fed in and allowed to
settle to the bottom. Each 3 to 4 cm of sediment was then compacted
underwater using a long tamping rod. The dry specimens were also
built up in layers and compacted in a similar manner.

After preparation of a specimen, a static axial load was
applied by placing weights on the platform at the bottom end of
the axial wire and by attaching upper inertia rings appropriate
for a particular test. Each specimen was allowed to consolidate
and adjust to the axial load level for a period of 24 h before
any dynamic loads were applied. A 24-h period of rest was also
allowed after any change in axial load for subsequent stages of
a test. This uniform rather lengthy time interval was adopted

because preliminary tests indicated that time-dependent effects often persisted for several hours after a change in load even in coarse materials.

Dynamic torsional excitation of the specimen was produced by placing one or more coils mounted on the top of the specimen in a magnetic field. In our early experiments with long, thin specimens, the field was produced by a large horseshoe magnet with special pole pieces, and an iron core was positioned inside the coil in the same manner as in a galvanometer with a D'Arsonal movement. The current flowing in the driver coil, and thus the frequency of torsional motion, was controlled by a variable frequency oscillator, a power amplifier, and a tone burst generator which was used as an electronic switch. In these earlier tests the rotational motion of the specimen was measured by a pair of horizontal accelerometers mounted at the outer edge of the top mass.

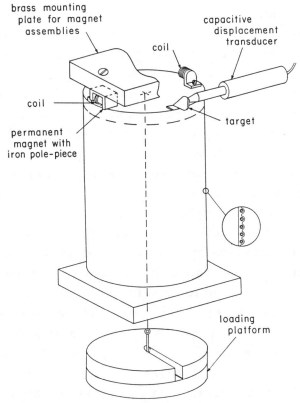

Fig. 4.3. Alternate setup to measure torsional log-arithmic decrement (Stoll, 1985).

In some of our more recent tests, paricularly those on shorter specimens of larger diameter, we have used pairs of coils and magnets located along the periphery of the upper mass as drivers (Fig. 4.3). In addition the motion of the specimen is measured with noncontacting capacitive displacement transducers that have the same sensitivity at all frequencies of interest. These transducers utilize a signal frequency of several mHz and so can be used to measure the displacement of a target at freqencies well over one kHz. Moreover, they can accurately resolve displacements of the order of 10^{-7}cm so that it is possible to monitor motion corresponding to maximum shear strains of less than 10^{-6} in the specimen. As we will see shortly, when strains rise above this level, the response of the sediment becomes nonlinear and the results are no longer representative of an acoustic wave propagating far from the source.

In each test the frequency was varied until a resonant mode of vibration was detected and for each mode the specimen was driven to resonance for a number of different power settings. At each resonance, the frequency was adjusted to obtain maximum amplitude after which the power was switched off electronically and the decay of the wave form recorded using a dual-beam storage oscilloscope or the data acquisition system of a microcomputer. A typical record showing the driver signal and output of one of the accelerometers is shown in Fig. 4.1. The logarithmic decrement is usually determined from the slope of a semilogarithmic plot of the amplitude for the first 8 or 10 cycles of the decay (see Richart, Hall and Woods, 1970, for a discussion of this technique). Six or seven power settings were applied for each resonant mode in order to investigate the influence of amplitude and the logarithmic decrement and resonant frequency. At each resonance, the low-amplitude response was investigated first in order to avoid disturbing the specimen, and the maximum amplitude was limited to a predetermined level to preclude any effects on subsequent measurements. Early tests showed that the damping characteristics of the specimen are temporarily changed by vibrations above a certain amplitude even though no tendency toward dilation or compaction is apparent.

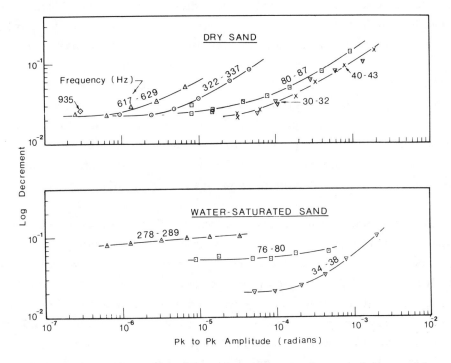

Fig. 4.4. Logarithmic decrement versus peak-to-peak rotation at top of specimen for "20-30" Ottawa sand with axial effective stress of 30kPa (Stoll, 1979).

Fig. 4.4 shows the results of a typical series of tests on dry and water-saturated Ottawa sand (sample no. 1) for a static axial load of 30 kPa. For a dry specimen the curves labeled 80-87, 322-337, 617-629, and 935 Hz represent the responses observed in the first, second, third and fourth modes , respectively, for the case where no added mass was attached to the top of the specimen (i.e., the top was composed of only a thin disk supporting the transducer and coil). The curves labeled 40-43 and 30-32 Hz represent the response with one and two brass rings added to the top assemblage. The total axial stress on the specimen remains unchanged when an inertia ring is added since a compensating weight is removed from the platform at the bottom of the axial loading wire.

In the case of water-saturated sand, the curves labeled 76-80 and 278-289 Hz are for the first and second mode with the minimum top mass, and the curve labeled 34-38 is for the case where one inertia ring has been added. The range of resonant frequencies noted above each curve gives some idea of the spread that was observed when the amplitude was varied during each run. In each case the higher frequency is the asymptotic value obtained at low amplitudes while the lower value corresponds to the largest amplitude used in a particular test. The latter value is somewhat arbitrary since the maximum amplitude was varied from test to test, depending on the power requirements and other factors. A discussion of the theory necessary to determine the mode shapes and to determine the maximum amplitude of shear strain from the motion observed at the top of the specimen is given in Appendix C.

For dry sand, the curves shown in Fig. 4.4 tend toward a constant log decrement at low amplitudes. The lowest point on each curve is about the limit of the experimental equipment because both low- and high-frequency noise tend to mask the signal at lower levels of excitation. For saturated sediment, the curve for the lowest frequency is similar to the curves observed for the dry sand with a definite tendency toward a constant decrement at low amplitudes. However, when the frequency is increased, there is a marked increase in log decrement for each step, and the curves change character completely becoming more nearly straight lines with no real asymptote. Thus the dominant effects of viscous losses in the fluid are clearly evident at the higher frequencies, whereas the frictional losses in the skeletal frame tend to control the attenuation at low frequencies. It should be noted that the low-frequency curve for saturated sand becomes asymptotic to about the same value of decrement as do all the curves for the dry sand. Thus it appears that the presence of interstitial water does not appreciably alter the magnitude of the frictional losses that occur at the grain contacts of the sediment.

If we use the data shown in Fig. 4.4 to construct a diagram of logarithmic decrement versus maximum cyclic shear strain, rather than amplitude of rotation at the top of the specimen, we obtain the results shown in Fig. 4.5. This figure clearly shows that there is a threshold of cyclic shear strain above which the damping in the dry sediment is a function of amplitude. In this case the threshold appears to be slightly above 10^{-6}. Below this

strain level the log decrement approaches a constant value of
2 to 3×10^{-2}. Hence we come to the very important conclusion that
only data obtained at levels below this threshold are valid for
use in a linear theory which assumes amplitude independence of
the complex moduli which govern the response of the skeletal
frame.

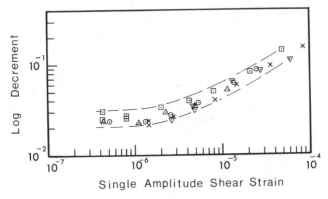

Fig. 4.5. Logarithmic decrement versus maximum cyclic
shear strain for dry "20-30" Ottawa sand with axial
effective stress of 30 kPa - same symbols as Fig. 4.4
(Stoll, 1979).

The real part of the complex shear modulus is also sensitive
to the amplitude of the maximum cyclic shear strain. In Fig.
4.3, the maximum resonant frequency for each mode, and therefore
the highest shear wave velocity, corresponds to the lowest shear
strain amplitude. This is completely consistent with the results
obtained by other investigators who have studied the response of
various sediments using several different experimental apparatus.
Figs. 4.6 and 4.7 show the results of some of this work. Fig.
4.6 is based on data from a series of experiments performed by
Iwasaki, Tatsuoka and Takagi (1977) who used cyclic shear and
conventional resonant column tests to study the response of dry
sand. The mean effective stress, $\sigma_0{}' = (\sigma_1{}' + \sigma_2{}' + \sigma_3{}')/3$, which is
the average of the three principal normal components of effective
stress, and porosity were held constant in this series of tests
so that the results reflect only the influence of the amplitude
of cyclic shear strain.

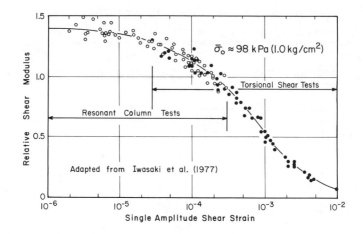

Fig. 4.6. Effect of strain amplitude on dynamic shear modulus (Stoll, 1980).

This remarkable set of data, which spans strains ranging over four orders of magnitude, clearly shows the influence of amplitude and the tendency for the dynamic shear modulus to approach a constant value at strains less than about 10^{-5}. It is important to note that the shear modulus is reduced by over 40% in going from a strain of 10^{-6} to strains of the order of 10^{-4}. Since the latter value is typical of the strain level used in many experiments reported in the geotechnical literature, care must be exercised in using these data to predict acoustic wave velocity where the strain amplitude is always very small.

Fig. 4.7, which illustrates the effect of strain amplitude on damping, has been adapted from data given by Seed and Idriss (1970). Their data were collected from a number of different sources so that the particle size, percent of different fractions, porosity and static stress are variable. However, even though the variation of these parameters leads to a considerable amount of scatter, the effect of cyclic strain amplitude and the tendency toward an asymptote at low amplitudes is still evident. The bounds of the data shown in Fig. 4.4 are also shown in Fig. 4.7 by the pair of broken lines.

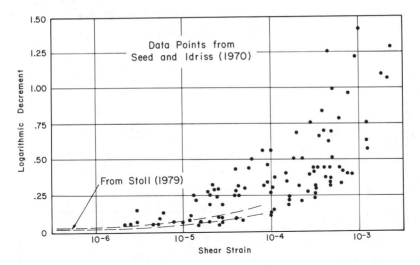

Fig. 4.7. Effect of strain amplitude on damping
(Stoll, 1980)

There are many other data in the literature, not only for sands but also for silts and clays, that suggest a behavior similar to that illustrated in Figs. 4.5 , 4.6 and 4.7. Thus it is important that a certain amount of care be exercised in planning experiments or extracting data from the literature if they are to be used in a linear model to describe the acoustic response of marine sediments. Only a small percentage of the available data has been obtained at amplitudes small enough to insure that it may be used directly in such a linear model.

Another set of data taken with the apparatus shown in Fig. 4.1 is presented in Fig. 4.8. These test results are for sample No. 2, which is a typical, naturally occuring beach sand with a response quite similar to that of Ottawa sand. Results for three different static stress levels are shown, and in all cases the log decrement was found to decrease with an increase in static stress level. This is consistent with the theoretical predictions given in Chap. 2 and with most of the experimental results in the literature (e.g., see Hardin, 1965b). This kind of information is extremely important in practical applications where the ambient or geostatic stress is continually varying with depth in the sea floor; unfortunately, in the series of tests depicted in Fig. 4.8, there were not enough data to assign an exact functional form to the manner in which the decrement changes with ambient stress. This problem will be considered in more detail in Chap. 5.

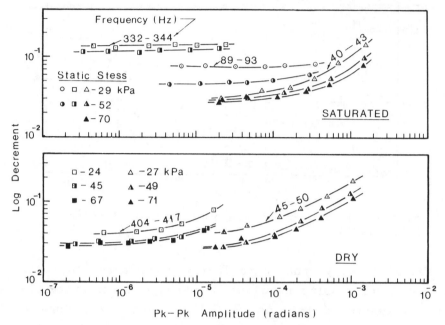

Fig. 4.8. Logarithmic decrement for Fire Island sand at various effective stress levels (Stoll, 1979).

DETERMINING 1/Q FROM PHASE MEASUREMENTS

The apparatus shown in Fig. 4.1 is limited to studies in the frequency range from 30 to about 350 Hz for saturated specimens. Since the range of interest in marine geoacoustic work runs from less than one Hz to about 10 kHz, we have developed new equipment which allows our studies of attenuation to cover a much wider frequency range (Stoll, 1984, 1985a and 1985). The experimental setup for the measurement of modulus and damping at very low frequencies is shown in Fig. 4.9. In these experiments the phase difference between a driving couple and the resulting torsional motion is used to measure the Q of the material. The experiments are performed with the aid of a microcomputer which is used to control the experiment and obtain robust samples of many cycles of loading. The microcomputer is also used to display and process data as well as to calculate the Fourier transforms that are used to determine the phase difference between the driving torque and resulting torsional motion.

The spirally reinforced confining shell used in these experiments is similar to the one described earlier except that the embedded filament is .25 mm nylon instead of stainless steel. Static axial load is still applied to the specimen by a thin steel wire running down the axis of the specimen and through a tiny gland to a loading platform situated beneath the bottom plate.

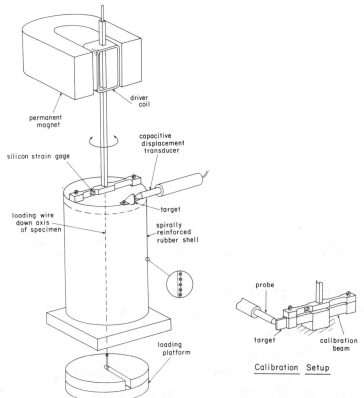

Fig. 4.9. Experimental setup for measurement of phase difference between driving torque and resulting motion of the top of a specimen (Stoll, 1985a).

Part of the success of this new apparatus was due to the availability of a commercial, noncontacting transducer that can resolve extremely small displacements. Such transducers are used commercially in manufacturing plants to sense the variation in thickness of thin sheets of different materials or to check the alignment of high-speed magnetic disks used for data storage in computers. The unit we are using generates a radio frequency signal (3mHz) at the tip of a probe positioned so that the measuring face is about .0025 mm from the moving target. The full scale output of the signal conditioner used with the probe is +/- .00025 mm, linear to within 0.4% of full scale output. Since we use the transducer in a range near its lower limit, extreme care in mounting the transducer and isolating the entire apparatus from external noise caused by air currents and structural vibrations is essential.

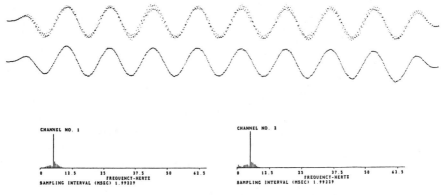

Forcing Function, Response and Spectra - 5 Hz

Fig. 4.10. Plot of data points and Fourier spectra for one window - both ends tapered (Stoll, 1985a)

The torque applied to the top of the specimen is measured by sensing the strain in a torque beam attached at two points to the top plate of the specimen. A silicon strain gage bridge and an instrumentation amplifier provide the necessary electrical output for sampling by an A/D converter. During the test alternate samples of the driving torque and resulting torsional motion are made and stored directly in the computer memory. In most cases a robust sample involving more than 100 cycle of motion is taken, after which a number of Fourier transforms are performed using a sliding window that is moved progressively through the data. A cosine taper is usually applied to 10% of the data at each end of the window. Fig. 4.10 shows a plot of the data points from one such window together with the amplitude spectra from an FFT. In addition to the FFT, a full FT was performed at frequencies near the maximum spectral ordinate. The phase of the two signals was then obtained by taking the ratio of the imaginary and real parts of the transform at the frequency corresponding to the maximum amplitude and correcting for the precise offset caused by the multiplexing.

Each phase difference measured in the manner described above must be corrected for any instrument error that may change the phase of the signal both before and after it is sampled. For example, a low-pass filter and other signal conditioning circuitry introduce some differential phase shifts no matter how carefully they are designed and matched. This is particularly important in view of the fact that we are attempting to measure real phase differences of the order of one degree or less! For this reason the entire system, operating at essentially the same frequency and amplitude as for the real experiments, was calibrated by substituting a solid aluminum beam for the specimen. The insert in Fig. 4.9 shows a schematic of the calibration setup. Since the aluminum beam has a Q at least one order of magnitude larger than the real specimen, one can use the response of the beam to determine the phase lag of the entire system and thus largely eliminate the instrument error. A schematic of the components used in these experiments is shown in Fig. 4.11.

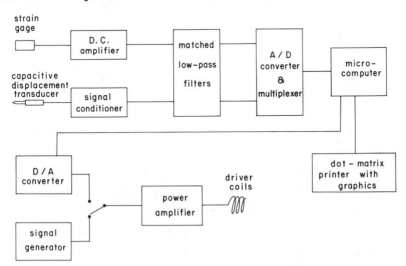

Fig. 4.11 Schematic of components used in experiments to measure phase difference between driving torque and resulting torsonal motion (Stoll, 1984).

     Results of the low-frequency experiments described above
were combined with the data obtained using the experimental setups
shown in Figs. 4.1 and 4.4 to obtain a continuous spectrum of
logarithmic decrements over the frequency range from one or two
Hz to well over one kHz for several different gradations of sand
and for an inorganic silt (No. 4 in Fig. 4.2). For example, for
20-30 Ottawa sand, we obtain the response shown in Fig. 4.12.
In this figure the results for dry sand show that the assumption
of a constant logarithmic decrement is quite reasonable at least
up to a frequency of a few kHz. Moreover, the fact that we are
able to document the constant decrement for frequencies up to a
kHz using several different kinds of equipment is a good check
on the reliability of our measuring techniques and confirms that
no spurious effects such as equipment resonances are affecting
the data in this range

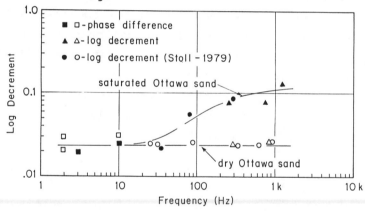

Fig. 4.12. Logarithmic decrement versus frequency for
water-saturated and dry 20-30 Ottawa sand (Stoll, 1985).

     When the sand is fully saturated, Fig. 4.12 shows that the
damping is strongly dependent on frequency as predicted by the
Biot theory (see Fig. 2.2), even though the same constant complex
moduli are used to describe both the wet and dry response of the
skeletal frame. At very low frequencies, the "overall" decrement
corresponding to both wet and dry sand is dominated by the damping
in the skeletal frame, and the values of the decrement fall in
the range of 0.02 to 0.03. This relatively low value for damping
is somewhat less than what was assumed in some of our early work
(Stoll and Bryan, 1970 and Stoll, 1974), which was based on

experimental data from early geotechnical literature where the
amplitudes of motion were not low enough to reach the quasilinear
range of response. In fact, the current values are similar to
the decrements measured for some lithified material such as weak
sandstone (Spencer, 1981, Murphy, 1984).

In order to investigate the effects of frequency on a saturated
sediment with somewhat lower fluid mobility we have performed an
extensive set of tests on an inorganic, micaceous silt with a
grain size curve as shown in Fig. 4.2, curve No. 4. Typical data
for the first and second modes of a resonant column test are shown
in Fig. 4.13 for three different levels of static confining
stress.

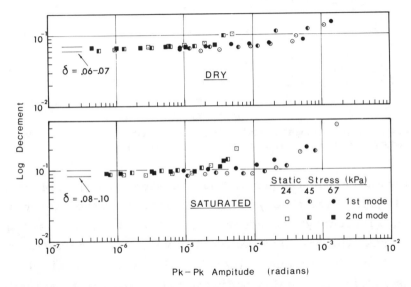

Fig. 4.13. Logarithmic decrement versus frequency for
water saturated and dry silt at different stress levels
(Stoll, 1990).

At different times during the period from 1979 to the present
we have performed tests on the same silt material using the
equipment shown in Figs. 4.1, 4.3 and 4.9. The results of this
series of tests are shown in Fig. 4.14. This figure contains the
same data shown in Fig. 2.6 with several new data points added
from recent tests. The data shown in this figure present a very
consistent picture of how the damping varies over the full range
from 2 Hz to over 1 kHz. There is a continuous increase in the
logarithmic decrement with increasing frequency up to the limit

of our data in the low kilohertz range. Above frequencies of
about 1.5 kHz resonances in our equipment began to affect the
data; however the trend of the data suggests that the values will
increase even more. We expect the maximum to occur somewhere in
the low kilohertz range analagous to the results for water-
saturated rock given by Spencer (1981) and Murphy (1982).

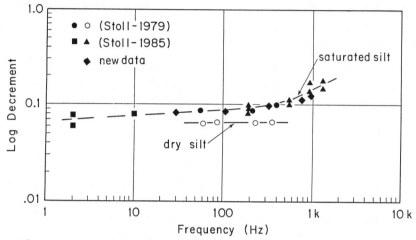

Fig. 4.14. Logarithmic decrement versus frequency for
water-saturated and dry silt (Stoll 1986).

The effect of frequency that is seen in Fig 4.14 cannot be
modeled by the Biot theory unless the moduli of the skeletal frame
are made frequency dependent. For the the silt used in our
experiments, the measured permeability was found to be 2.5×10^{-10}
cm^2 at a porosity of approximately 50%. When a constant complex
modulus is used to model a sediment with this low a permeability,
the theory does not predict any appreciable increase in log
decrement until the frequency is over 10 kHz (see Fig. 2.7).
Thus, the frequency dependence of Q that is obvious from our
experiments must be incorporated into the functional form of the
moduli used to describe the response of the skeletal frame in a
water environment. It should be noted that this is not the
equivalent of modeling the overall response of the sediment on
the basis of a single set of viscoelastic parameters because of
the coupling between the motion of the skeletal frame and the
overall fluid motion relative to the frame which is taken into
account by the Biot theory.

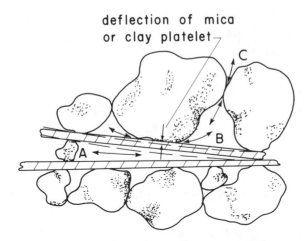

deflection of mica
or clay platelet

C

B

A

Fig. 4.15. Schematic of sediment showing regions
A, B and C where "squeeze-film" motion may occur.

For the low-amplitude waves which we wish to model, a linear
viscoelastic model appears to be appropriate for the purpose of
describing the response of the frame in a water environment. As
pointed out earlier, we attribute the damping at lower frequencies
in the finer sediments to be the result of local fluid or "squeeze
film" motion caused by relative approach of the particles near
various intergrain contacts. The spectrum of relaxation times
associated with this process will depend on the complexity and
variability of the actual structure in the many different kinds
of marine sediment. Fig. 4.15 is a sketch which shows various
kinds of local fluid motion that would contribute to the viscoelstic
response in a real sediment. The form of viscoelastic model that
can be used to fit the kind of data shown in Fig. 4.14 will be
discussed in Chap. 5.

RECENT EXPERIMENTS TO RELATE G AND K

In order to use any version of either the Biot or Gassmann
models to predict the propagation characteristics of p-waves, it
is necessary to know the relationship between any two complex
moduli that are to be used to describe the response of the skeletal
frame. In the laboratory it is generally the complex shear modulus

$\overline{\mu}$ and the complex Young's modulus $\overline{E}$ that are measured. Because of the fixed diameter and length of laboratory specimens, measurements are often made at a few discrete frequencies that are dictated by the various resonances of the sample. Since the resonant frequencies of shear and extensional modes do not generally coincide, it is a difficult task to establish experimentally the relationship between the real and imaginary parts of $\overline{\mu}$ and $\overline{E}$ if the moduli exhibit any appreciable frequency dependence. Moreover, when the sediment is fully saturated, the results of experiments utilizing the extensional mode may depend on the radial boundary conditions (i.e., open or closed - see Dunn (1988)) and since both dilatational and shearing motion are occurring simultaneously, the interpretation of experimental data is not simple.

The ideal experiment would allow the measurement of both $\overline{E}$ and $\overline{\mu}$ over an extended frequency range on the same specimen. As shown in Figs. 4.12 and 4.14 we have been able to accomplish such measurements over an extended frequency range for the case of shearing motion, using resonance techniques in the higher frequency range and relative phase measurements at low frequencies. Unfortunately, the reinforced shell and our method of static loading, which are ideal for studying shearing motion, have drawbacks when they are used in studies of extensional motion. The reinforced shell applies a varying radial traction to the peripheral boundary of the specimen depending on the amount of radial displacement. In addition, the effect of the axial loading wire must be taken into account in evaluating the axial stiffness. While corrections can be made to compensate for both of these effects, the results are not as "clean" as those obtained for the case of torsional shearing motion. A schematic diagram showing the various tractions and stiffness components of the apparatus are shown in Fig. 4.16.

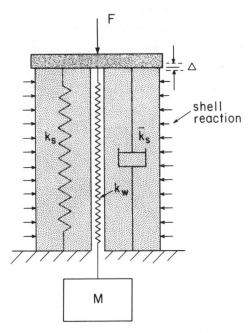

Fig. 4.16. Schematic diagram showing various stiffness components and boundary conditions on our experimental configuration.

The basic moduli input to both the Biot and Gassmann models are the complex bulk modulus, $\overline{K}$, and the complex shear modulus, $\overline{\mu}$, of the skeletal frame determined in a water environment under conditions where the the pore water pressure remains constant. Thus in the laboratory we need to measure the drained, "jacketed" compressibility of a water-filled specimen to obtain the appropriate complex bulk modulus, or alternatively, to measure some other modulus such as $\overline{E}$ under the same drained conditions and use both $\overline{E} = E_r + iE_i$ and $\overline{\mu} = \mu_r + i\mu_i$ to calculate $\overline{K}$ using

$$K_r = (\mu_r + E_r)/3(3\mu_r - E_r) \tag{4.1}$$

$$K_i = (3\mu_r^2 E_i - E_r^2 \mu_i)/3(3\mu_r - E_r)$$

Unfortunately measurements of $\overline{E}$ or $\overline{K}$ involve volume change in the specimen and therefore the compressibility and mobility of the fluid will play a role in the interpretation of the experimental results (i.e., the full theory must be used) if undrained dynamic tests are employed. One alternative which we have undertaken is to run a few tests on partially saturated specimens so that the effect of fluid compressibility is suppressed.   Under these conditions local viscous effects still influence the overall response although possibly in a slightly modified manner because of capillary effects.   The capillary effects would become increasingly important in tests of this kind on very fine soils such as clay.

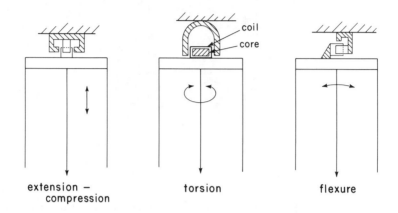

extension — compression   torsion   flexure

Fig. 4.17. Schematic showing three different kinds of electromagnetic driver used to excite specimen.

A third kind of laboratory test involving flexural motion is also often used to study the complex moduli of various anelastic materials.   We have modified the electromagnetic driver used to excite long slender specimens of the kind shown in Fig. 4.1 in order to produce flexural motion in addition to the other two modes (torsion and compression-extension).   A schematic showing the three different driver configurations is shown in Fig. 4.17. Again, there are problems involved when testing specimens of sediment whether it be in our special shell or in the standard resonant column devices that are sold commercially.   Flexural experiments work well with very long, thin rods where the deformation is controlled mainly by Young's modulus, $\overline{E}$ and the effects of shear and rotatory inertia are small enough to be neglected.   However, as the ratio of diameter to length of a specimen increases to a value of about 0.1 or larger, these effects become important and the analysis becomes much more complex since both dilatational and shear strains play a role in determining the mode shapes. However, the results from Timoshenko beam theory and other higher order solutions are available in the literaure in a form that can be used to help in the interpretation of this kind of experiment (i.e., see Blevins, 1979, Stokey, 1976, Spinner and Tefft, 1961).

We have performed a few tests involving axial and flexural motion on specimens of the silt sediment described previously and the results are shown in Fig. 4.18 superimposed on the data previously given in Fig. 4.14. Both the observed response and the data corrected for the effects of our experimental configuration are shown in this figure. While there are only a few data points, they follow quite closely the trend of the data from all of our torsional experiments suggesting that, until we have more data, a reasonable assumption may be to take the log decrement corresponding to $\bar{E}$ as equal or slightly larger than the decrement for $\bar{G}$, with the same frequency dependence. Clearly this is an area of experimental research that needs further work.

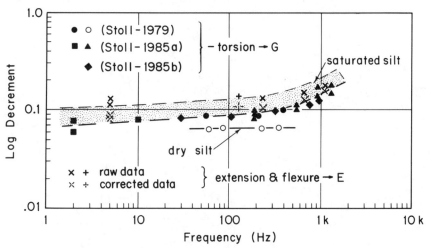

Fig. 4.18. Logarithmic decrements corresponding to both $\bar{G}$ (or $\bar{\mu}$) and $\bar{E}$ from torsional, flexural and compressional-extensional tests (Stoll, 1990).

From the laboratory data summarized in this chapter, it is clear that the assumption of a constant logarithmic decrement or Q, or equivalently an attenuation constant α that depends on the first power of frequency, is unacceptable in the case of nearly all marine sediments. For coarse sediments such as sand the fluid

mobility predicted by the Biot theory produces viscous losses
that render the overall response frequency dependent. In finer
sediments there are viscous losses associated with local fluid
motion that is intimately related to the deformations of the
skeletal frame near the points of intergranular contact. Since
these two forms of damping are fundamentally different and since
they may occur simultaneously, the overall response cannot be
described by a simple viscoelastic model controlled by a single
relaxation function. The ability to incorporate both of these
rate dependent processes into the fundamental Biot theory, by
associating the local losses with the complex moduli of the
skeletal frame and the overall relative motion of the fluid field
with Biot's viscodynamic operator, demonstrates the power and
flexibility of his theory.

CHAPTER 5

EVALUATING THE BIOT PARAMETERS

The basic parameters of the Biot theory, as augmented for geoacoustic modeling, can be divided more or less into two categories - the passive constants such as density of the various components, the porosity and the compressibility of the solid and fluid components and a second category which includes a number of interrelated parameters that define the two frequency-dependent operators governing the relative motion of the fluid with respect to the skeletal frame and the viscoelastic response of the skeletal frame. In this chapter we will concentrate mainly on the parameters in the second category, utilizing the concepts discussed in the first four chapters and various data from the literature.

BIOT'S VISCODYNAMIC OPERATOR

The viscous losses associated with the overall movement of the pore water relative to the skeletal frame are included in the Biot model via an operator that involves the first and second partial derivatives with respect to time of the variables ζ and $\vec{\theta}$ (i.e., see Eqs. 1.17 and 1.23).

$$\left(m\frac{\partial^2}{\partial t^2} + \frac{\eta}{k}\frac{\partial}{\partial t} \right)(\zeta \text{ or } \vec{\theta}) \tag{5.1}$$

In the simultaneous solution of either Eqs. 1.15 and 1.17 or Eqs. 1.23, Biot introduces a "complex correction factor" which is applied in the frequency domain to correct for deviations from Poiseuille flow that occur as the frequency is increased. Biot's correction factor is based on the actual microvelocity field in a system of highly idealized "channels" that are meant to represent the paths of flow through the porous medium. In particulate materials, such as ocean sediments, it is particularly difficult to reconcile such a simple model with the tortuous interparticle geometry that actually prevails. Not only will there be a deviation from laminar or Poiseuille flow in the ducts of uniform size and shape treated by Biot, but also abrupt changes in direction and the expansion and contraction of the flow channels. Thus, rather than trying to model the actual microvelocity fields in an almost infinite variety of different sediments, we have treated the parameters a, α, and k (Eqs. 1.18 and 1.19) as phenomenological variables to be determined from experiment rather than theoretically derived from the microgeometry of the pores.

The parametric studies given in Chap. 2 help to sort out the relative effects of each of the variables in question. For example, it is clear from Figs. 2.1 and 2.2 that there is a "characteristic" frequency at which the maximum amount of intrinsic damping occurs associated with each value of k/a^2 and that the effect of decreasing the permeability is to shift this frequency to higher and higher values. In fact, the characteristic frequency shifts about one decade in value for each decade of decrease in k. Since the coefficient of permeability for naturally occurring sediments tends to range over 7 or 8 orders of magnitude (Lambe and Whitman, 1970), it is clear that the particle size and permeability play an important role in determining when this kind of damping will play an important role in the overall response of a sediment.

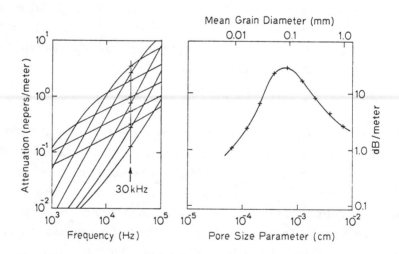

Fig. 5.1. Relationship between attenuation and grain size (Stoll, 1974).

In order to relate the permeability and the characteristic frequency to different kinds of sediment, we have made use of attenuation curves similar to those shown in Fig. 2.4. At a fixed frequency, say 30 kHz, a curve relating attenuation to pore size parameter may be derived as shown in Fig. 5.1. This bell-shaped curve is similar to experimental curves (also for 30 kHz) presented by Shumway (1960), McCann and McCann (1969), and Hamilton (1980). Hamilton's collection of data is shown in Fig. 5.2. If the peak of the theoretical curve shown in Fig. 5.1 is compared with peak of the experimental curves, a correspondence between the pore size parameter, α, and the mean grain diameter may be obtained. The result is the upper scale of mean grain diameter shown in Fig. 5.1. For the particular set of parameters used to construct the family of curves shown on the left side of Fig. 5.1, it appears that a reasonable choice for the pore size parameter is a value between 1/6 and 1/7 of the mean grain diameter.

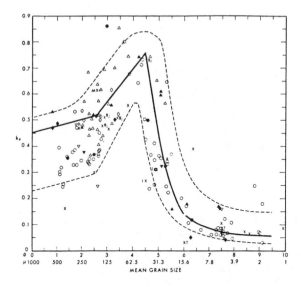

Fig. 5.2. Attenuation versus grain size (Hamilton, 1980).

The third of Eqs. 1.19, as well as Figs. 1.3 and 2.5, show that the pore size parameter acts as a scaling factor which determines the relative sharpness of the peak that corresponds to the characteristic frequency in the damping curves. The sharpness of these curves in turn influences the shape and slope

of the attenuation curves in the high-frequency range as illustrated in Fig. 2.6.   The other free parameter, α (or m), tends
to shift the attenuation curves vertically without much change
in shape as shown in Fig. 2.7.   Thus while it is clear that all
three of the parameters k, α, and a are interrelated and, as a
result, limited in their range of values we still have considerable
leeway in choosing a reasonable set that will match the observed
response of real sediments.

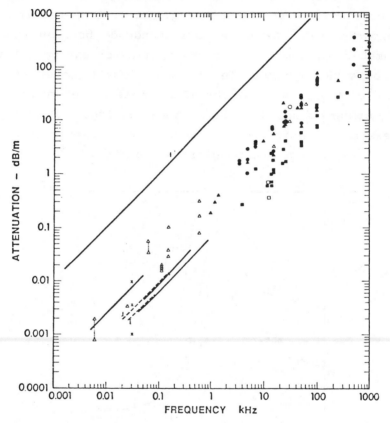

Fig. 5.3.   Global plot of p-wave attenuation versus
frequency (Hamilton, 1980).

Several investigators,including Johnson, Koplik and Dashen
(1987) and Yamamoto and Turgut (1988), have proposed alternate
descriptions of "dynamic permeability" that are intended to account
for the effects of the complicated pore geometry in real sediments.
Johnson et al. define a modified function $F(\omega)$ (e. g., see $F(\kappa)$,
Eq. 1.19) that depends on a new parameter related to the geometry
of the sediment interstices (surface area to pore volume ratio).

Yamamoto and Turgut propose yet another function by assuming that the pores can be modeled by a bundle of capillaries with a distribution of radii related to the pore-size distribution of the sediment. Both of these theories are appealing in that they attempt to account for the complicated pore geometry of real sediments; however, both require additional quantitative information about the pore-size distribution which may be difficult to provide in many practical applications.

Since one of our primary goals has been to develop a theory that will provide a satisfactory link between the attenuation observed at high frequencies and the damping that is measured at low frequencies, it is important to examine the actual measured values of attenuation for different classes of sediment. The historical tendency has been to lump all of the attenuation values for sands, silts and clays into one global plot and then infer the frequency dependence of attenuation from the overall trend. One such collection of data has been given in a number of papers by Hamilton (1972, 1980). One version of Hamilton's plot is shown in Fig. 5.3. Much of the data in this figure was taken at frequencies over 1 kHz since it is much easier to obtain lab and field data relating to attenuation in this range. In order to differentiate between the response of "coarse" sediments such as sand and the finer sediments such as clay and silt, it is helpful to examine the data from each category seperately with enough detail to permit examination of the frequency dependence for each of the experiments that are included in the two categories.

Fig. 5.4 shows a collection of data for sands while Fig. 5.5 shows the attenuation for finer material all of which contain less than 1% sand. A "match line" with a slope of unity has been superimposed on each figure in order to make it easy to visualize the relative position of the data in the two figures as well as to permit an estimate of the slope of the data sets from each of the experiments. Most of the data contained in Hamilton's global plot, Fig. 5.3, is included in Figs. 5.4 and 5.5. A comparison of these three figures shows that the attenuation observed in the coarser sediments is significantly higher than for the clays and silts; as a result, nearly all of the data for the sands falls in a cluster near the upper bound of the Hamilton

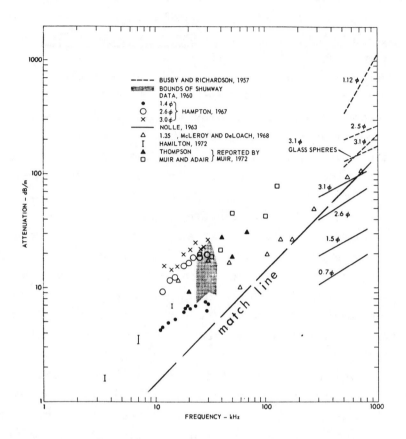

Fig. 5.4. Attenuation versus frequency for sand
(Hampton and Anderson, 1974).

plot whereas the data for silts and clays fall in the lower half
of the data band. Moreover, the trends of the data from most of
the individual experiments on sand suggest that the attenuation
is varying as the frequency to some power significantly less than
one (i.e., slope of attenuation curve on log-log plot < 1).

For reasonable choices of the variables k, a and α, the kind
of response described above is precisely what the Biot theory
predicts. In the case of sands, at very low frequencies, the
theory suggests that the attenuation will fall significantly below
the Hamilton data band (Fig. 5.3) whereas in the high frequency
range the predicted attenuation falls near the top of the data
band. At the low frequencies, the position of the attenuation
curve depends on the logarithmic decrement assigned to the skeletal

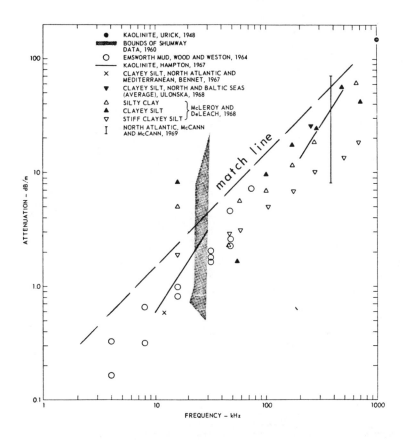

Fig. 5.5.   Attenuation versus frequency for clays
and silts - less than 1% sand (Hampton and Anderson
1974).

frame, whereas in the high-frequncy range the fluid mobility
becomes increasingly important and viscous losses tend to dominate
the observed overall response.

In Fig. 5.6 we have added some additional data to Hamilton's
collection and superimposed some predictions of the model for
comparison.   The theoretical curves labeled "SAND" start below
the Hamilton band and end up near the upper edge of the band in
the high-frequency range.   Thus the theoretical curves provide
what appears to be a reasonable link between the high- and
low-frequency measurements.   The heavily shaded band labeled
"SILT" is based on the experimentally determined variation of
logarithmic decrement shown in Fig. 4.18.   From the figure it can

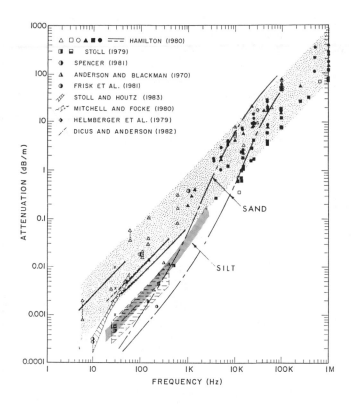

Fig. 5.6.   Laboratory and field data for attenuation
versus frequency with curves showing the predictions
of the model superimposed (Stoll, 1986).

be seen that the model predicts values of attenuation approaching
the lower portion of the Hamilton data band as the frequency
increases to 1 kHz or more.   Again the model predicts the correct
transition between the older high-frequency data and the new data
obtained at low frequencies; however, the mechanism in this case
is viscous damping associated with deformation of the skeletal
frame and not overall motion of the fluid as for the sands.   In
a subsequent section of this chapter (i. e., FRAME DAMPING-
VISCOELASTIC MODELING) we fit a viscoelastic model to this data
and incorporate it into the model so that the overall attenuation
is automatically calculated at any frequency.

SKELETAL FRAME PARAMETERS

The complex moduli, $\overline{K}$ and $\overline{\mu}$, which determine the response of the skeletal frame in a water environment are in general frequency-dependent parameters that depend strongly on the mean intergranular or "effective" stress and on the voids ratio or porosity. The theoretical discussions given in Chap. 3 and many experimental investigations suggest that the functional form of the real part of the dynamic shear modulus should be

$$\mu = f(e) \cdot (\sigma_0')^n \tag{5.2}$$

where

$$\sigma_0' = (\sigma_1' + \sigma_2' + \sigma_3')/3$$

is the mean effective stress and n is a power typically in the range of .33 to .67. The function f(e) has traditionally been chosen to fit a particular data set usually associated with one type of sediment over a limited range of voids ratio. For example, in early studies of the dynamic response of soils, Hardin (1965a) and Hardin and Richart (1963) developed the formula

$$\mu = \frac{1230(2.97 - e)^2}{1 + e}(\sigma_0')^{1/2} \tag{5.3}$$

for angular sands (the units of μ and σ_0' in Eq. 5.3 are pounds per square inch). A number of subsequent investigations have verified the usefulness of this equation for other particulate materials such as silt and clay, provided the numerical constants are slightly modified (Hardin and Black, 1968). Thus, in order to use this equation, different forms of f(e) are required for different strata if the type of sediment changes appreciably. In an effort to find a more general form for f(e) that might be used as a first approximation over a wide range of sediment types, Bryan and Stoll (1988) recently proposed a functional form for f(e) such that

$$\mu / p_a = a \cdot \exp(b \cdot e) \cdot (\sigma_0' / p_a)^n \tag{5.4}$$

where p_a, the atmospheric pressure, is introduced to permit the use of nondimensional arguments. a, b and n were chosen on the basis of a regression analysis on e and σ_0' from a population of

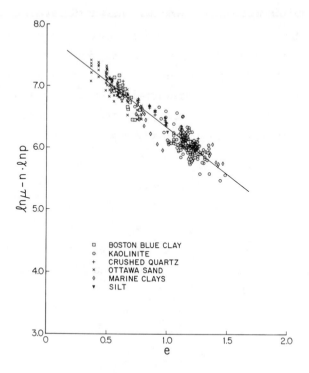

Fig. 5.7.   Results of regression analysis - e varying
(Bryan and Stoll. 1988)

test results involving sands and clays of low plasticity, silts
and various marine sediments. The regression on 494 points yielded
the following values:

$$a = 2526, \qquad b = -1.504, \qquad n = .448$$

with a correlation coefficient of 0.971. The results of the
linear multiple regression of $\ln(\mu/p_a)$ on $\ln(p/p_a)$ and e are shown
in Figs. 5.7 and 5.8.

The results shown above suggest that a single set of regression
coefficients can adequately represent a wide variety of sediment
types over the range of voids ratio and overburden pressure
typically found in marine sediments. Thus Eq. 5.4 can be viewed
as a first order, material-independent theory that is useful in
predicting the laboratory response of unconsolidated sediments
where the sediment type is not very well known. It is important
to remember that the data fitted to Eq. 5.4 are based on laboratory

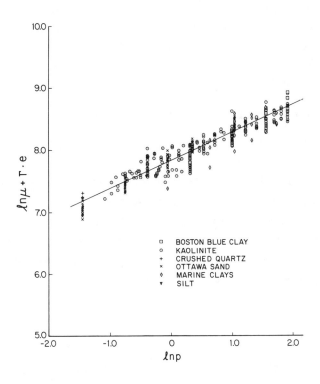

Fig. 5.8. Results of regression analysis - p varying
(Bryan and Stoll, 1988)

studies and that the in situ response of these same sediments
will vary somewhat depending on a number of different factors.
These include the effect of aging, stress history (i.e., overconsolidation ratio), cyclic shear strain amplitude and sampling
disturbance. All of these factors tend to increase the dynamic
shear modulus of an in situ sample as compared to laboratory
values that are often obtained at somewhat higher strain levels
after a short period of consolidation. Various comparative studies
have suggested that the increase will cause the value of a to be
in the range of 1.3 to 2.5 times the values obtained in the lab
(Arango, Moriwki and Brown, 1978, Anderson, Espana and McLamore,
1978). The differences between laboratory and field values for
various parameters will be discussed in Chap. 6 wherein some
recent field experiments are described.

The relationship between the real parts of $\overline{K}$ and $\overline{\mu}$ may be estimated by comparing the velocity of p-waves and s-waves in dry granular materials. In an elastic material the ratio of p-wave velocity to s-wave velocity is given by

$$R = v_p / v_s = \left(\frac{\lambda + 2\mu}{\mu}\right)^{1/2} = \left(\frac{2(1-\nu)}{1-2\nu}\right)^{1/2} \qquad (5.4)$$

and thus Poisson's ratio, ν, may be obtained from

$$\nu = \frac{2 - R^2}{2(1 + R^2)} \qquad (5.5)$$

Tests on granular materials suggest that the value of R and therefore ν are essentially independent of confining pressure at relatively low geostatic stress levels, with values for ν in the range of 0.1 to 0.2 being typical. Thus we would expect the real parts of both $\overline{K}$ and $\overline{\mu}$ to follow a pressure and voids ratio dependence such as that given by Eq. 5.4 with

$$K = \frac{2\mu(1 + \nu)}{3(1 - 2\nu)} \qquad (5.6)$$

approximately.

FRAME DAMPING - VISCOELASTIC MODELING

The amount of damping assigned to the skeletal frame and its frequency dependence depends on the coarseness of the sediment. In coarse-grained sediments such as sands we have found that a constant complex modulus is sufficient to model the frictional losses that occur at the grain contacts. However, in finer sediments such as silts and clays, local viscous losses attributed to "squeeze film" phenomena near each of the intergrain contacts require that the complex moduli be made frequency dependent.

In the coarser materials we have found that the logarithmic decrement for shear strain levels approaching those to be expected in normal seismic and acoustic signals far from the source (i.e., $< 10^{-6}$) are quite small, typically approaching value about equal to those measured in unconfined sedimentary rocks of relatively high porosity. For example, if we look at the data presented in Figs. 4.4 and 4.8, the asymptotic values of the logarithmic decrement at low strain levels is in the range of 1 to 3 times 10^{-2}. When these low values are used in the model, the result is that the attenuation of both shear and dilatational waves is very small at low frequencies. In fact the predicted values fall significantly below the "Hamilton" data band shown in Fig. 5.3 and 5.6.

Overburden pressure is another important factor to be considered in studies of intrinsic attenuation. In his studies of damping in sands, Hardin (1965b) found that the logarithmic decrement for torsional vibrations was proportional to $(\sigma_0)^{-0.5}$, whereas in our discussion of contact mechanics in Chap. 3 it was pointed out that the decrement should vary as mean pressure to the minus 2/3 power. Some very recent field experiments with shear and interface waves (Stoll et al, 1988) tend to verify this kind of response and to illustrate the very large changes that may occur very near the sediment-water interface as a result of the variation in overburden pressure. These results will be discussed further in Chap. 6.

In order to incorporate into the Biot model the kind of frequency dependence shown in Figs. 2.7, 2.8 and 4.18, we employ a viscoelastic model that can be used to describe the frequency dependent response of the skeletal frame in a water environment owing to the effects of local fluid motion. Squeeze film theory suggests that the additional intergranular force between particles caused by relative motion between particles of radii a and b is of the form

$$F = \frac{6\pi v}{h(a^{-1}+b^{-1})^2} \frac{dh}{dt}$$

(5.7)

$$\rho \frac{h}{v} \frac{dh}{dt} \ll 1$$

where h is the particle separation near the point of closest approach, v is the fluid viscosity and ρ is the mass density of the fluid. In order to assess the influence of intergranular viscous forces, it is instructive to consider a very simple model composed of a regular array of particles as was done in Chap. 3. If we choose a simple cubic array and assume that the particles are essentially spheres of equal radius, with small surface aberrations or small foreign particles causing a slight separation as in Fig. 5.9, then Eq. 5.7 can be used to predict the response for loading in a principal direction. The Hertz theory indicates that when two convex elastic bodies are in contact, most of the strain that results in relative approach of the two particles occurs in a small volume of each particle in the immediate vicinity of the small contact area. Thus, an assemblage of particles can be modeled approximately as an array of rigid bodies separated by small, nonlinear "springs". If we consider a stress field producing normal forces in a principal direction (i.e., parallel to a line connecting the centers of the spheres), then using Eq. 5.7 we may derive the following stress-strain relationship

$$\bar{\sigma} = M\epsilon + M' \cdot \frac{d\epsilon}{dt} \tag{5.8}$$

where $M' = (3\pi v/8)(D/h)$, D is the average particle diameter, ϵ is strain and M is a modulus that accounts for the spring constants at the grain contacts plus the incompressibility of the fluid between the grains. The effective stress, $\bar{\sigma}$, is the total force transmitted across a representative area of the assemblage divided by this area.

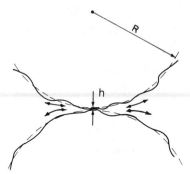

Fig. 5.9. Schematic of sediment grains showing local fluid motion (Stoll, 1980).

Eq. 5.8 describes the response of a simple viscoelastic material if M and M' are assumed to be independent of dynamic strain amplitude. This is a reasonable assumption if small oscillatory motions about some initial static strain level are considered and

if it is specified that M and M' depend on the static stress
level. Carrying the analysis one step further, we assume harmonic
motion and calculate the logarithmic decrement

$$\delta = \pi (M' \omega / M) \qquad\qquad (5.9)$$

where ω is circular frequency. For small strains, the log
decrement is related to attenuation α by the expression
$\alpha = \delta \cdot$ (frequency/phase velocity). For dry sediments such as sand
and silt, the log decrement generally will be in the range of
10^{-2} to 10^{-1} when friction is the only damping mechanism. If a
value of M typical of granular sediments (e.g., 6×10^7 Pa) is
substituted into Eq. 5.9 and various frequencies are considered,
the ratio of D/h that is necessary to produce the same amount of
damping that is caused by friction may be calculated. For example,
at 1 kHz, the above value of M leads to a value of D/h between
10^{-4} and 10^{-5}; each decade of increase in frequency has the effect
of decreasing this ratio by a factor of 10.

In real sediments where the particle size and shape may be
quite variable and where contact forces will have both normal
and tangential components, other components of local motion will
occur near the grain contacts. These components can also be
analyzed using lubrication theory. When all of the components
are combined, the result is a viscoelastic response similar to
that described by Eq. 5.8 except that M' will be a complicated
function that depends on the spectrum of particle sizes and shapes
and on the packing of the sediment.

In the application of our model to real sediments we will
choose M and M' in the simplest form that is compatible with our
experimental results. Since M represents the stiffness and damping
due to contacts between particles in a water environment without
local viscous losses, we cast it as a complex constant with the
ratio of imaginary to real parts determined by the amount of
damping that is observed at very low frequencies where the contribution of M' is very small. On the other hand, the form of
M' will be dictated by the frequency-dependent response that
occurs in the higher frequency range. Thus we choose M' to be
a viscoelastic operator with a spectrum of relaxation times based
on our experimental results. For example, if we wish to model
the kind of response shown in Fig. 4.18, any number of different

viscoelastic models are suitable; as an example we will use the Cole-Cole model employed by Spencer (1981) to describe the response of waterbearing sedimentary rock. However, we apply it in a somewhat different manner than Spencer in that we wish to model M' only and not the overall response of the sediment.

The real and imaginary parts of the Cole-Cole model, N_r and N_i respectively, are given by

$$N_r - N_\infty = \frac{(N_0 - N_\infty)\left[1 + (\omega\tau_0)^{1-\alpha}\sin\frac{1}{2}\alpha\pi\right]}{1 + 2(\omega\tau_0)^{1-\alpha}\sin\frac{1}{2}\alpha\pi + (\omega\tau_0)^{2(1-\alpha)}}$$

$$N_i = \frac{-(N_0 - N_\infty)(\omega\tau_0)^{1-\alpha}\cos\frac{1}{2}\alpha\pi}{1 + 2(\omega\tau_0)^{1-\alpha}\sin\frac{1}{2}\alpha\pi + (\omega\tau_0)^{2(1-\alpha)}} \qquad (5.10)$$

where N_0 and N_∞ are the relaxed and unrelaxed elastic moduli, τ_0^{-1} is the frequency at which N_i is a maximum and α is a parameter which determines the sharpness of resonance of the response curves. The distribution of relaxation times corresponding to the complex modulus $\overline{N} = N_r + iN_i$ is given by

$$F(s)ds = \frac{1}{2\pi} \cdot \frac{\sin\alpha\pi\,ds}{\cosh(1-\alpha)s - \cos\alpha\pi} \qquad (5.11)$$

In the above equations, $\alpha = 0$ corresponds to a single relaxation and therefore a distribution of relaxation times that is a delta function, whereas α greater than about .4 corresponds to a broad distribution of relaxation times. Plots of $1/Q = N_i/N_r$ versus frequency and the relaxation spectra for several values of α are shown in Fig. 5.10. In our model we choose N_0 to be zero since there will be a vanishing influence of local viscous losses as the frequency approaches zero. This leaves three free parameters, α, τ_0 and N_∞ which may be chosen to fit our experimental data.

It should be noted that when we specify the complex modulus in shear or compression to be the sum of a constant complex part plus a part defined by Eq. 5.10, we are creating an "overall" model that does not satisfy the Kramers-Kronig relations that are necessary for causality even though the Cole-Cole model by itself is in accord with this criterion. The same is obviously true when we use the constant complex modulus alone for coarser sediments. However, as pointed out in our first paper (Stoll and Bryan, 1970), it is possible to construct viscoelastic models

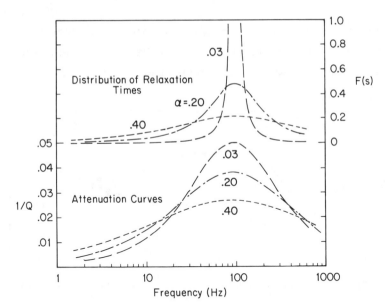

Fig. 5.10. Distributions of relaxation times and 1/Q for different values of α (Spencer, 1981).

with rather complex distributions of relaxation times which will satisfy the requirement of causality and also result in a response which is equivalent to a constant modulus over an extended but finite frequency range. However, as an expedient alternative we use the constant complex modulus together with a relatively simple form of relaxation spectra to produce a simple, useable model valid over a limited frequency range realizing that we are, in fact, approximating a slightly nonlinear behavior.

Thus $\bar{\mu}$ and $\bar{K}$ are chosen to be the sum of a constant complex modulus, say $\bar{\mu}_1$, plus a frequency-dependent complex modulus $\bar{\mu}_2$ determined from Eq. 5.10 so that

$$\bar{\mu} = \bar{\mu}_1 + \bar{\mu}_2$$

An example of such a model fitted to the data of Fig. 4.18 is shown in Fig. 5.11. Since our data do not extend above about 1.5 kHz, the location and amplitude of the maximum logarithmic decrement are rather arbitrary. Many other combinations of $\alpha, \tau_0^{-1},$ and μ_∞ will undoubtedly fit the limited data set equally well so that more data from the high-frequency range is necessary in order to make the best final choice. Nevertheless, with the

values used in the example we are able to describe $\bar{\mu}$ quite
satisfactorily in the range from 10 Hz to 1.5 kHz which is of
prime interest in geoacoustic modeling.

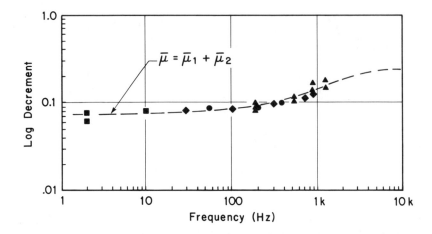

Fig. 5.11. Fit of constant complex modulus plus
Cole-Cole Model to data for silt (Stoll, 1990).

When the overall complex shear modulus as determined above
is introduced into the Biot model, with the logarithmic decrement
for shear and dilatation assumed to be equal, we obtain response
curves such as those shown in Fig. 5.12 labeled "constant complex
modulus + viscoelastic losses". In contrast, the curve labeled
"constant complex modulus" is the response that would have been
predicted had we chosen the complex modulus of the skeletal frame
to be constant thus neglecting the effects of local fluid losses
but including frictional losses and the usual viscodynamic losses
resulting from overall motion of the pore water relative to the
frame. In the latter case we see that the logarithmic decrement
is nearly constant until we approach frequencies of 2 to 3 kHz
in the case of dilatational waves and and even higher for the
case of shear waves. Thus in the low-frequency range it is the
local viscous losses together with the intergranular friction
that dominate the overall response and the effects of global
relative fluid motion do not become important until waves of
higher frequency are encountered. In finer materials of lower
permeability such as clay, this effect will be shifted to an even
higher frequency range. Thus, in the lower frequency ranges of

interest in most seismo-acoustic studies, a reasonable model for
the viscoelastic response of the skeletal frame in a water
environment is very important.

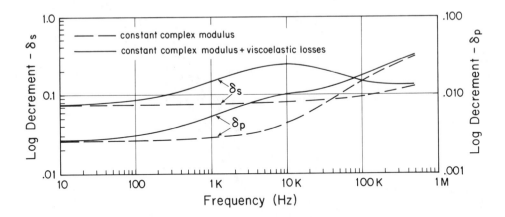

Fig. 5.12. Overall response based on Biot theory with
and without effects of "local" viscous losses
(Stoll, 1990).

When the attenuation of dilatational waves is calculated for
the general case, using the Cole-Cole model to account for the
local viscous losses, we obtain the result shown in Fig. 5.13.
In this figure the cross-hatched band labeled "SILT" corresponds
to a range of values for δ_μ/δ_E of 1 to 1.5 as suggested by our
experimental data (see Fig. 4.18). From this figure it is clear
that the predictions of the model, based entirely on the laboratory
results described previously, agree quite well with published
data over a wide frequency range. In the lower frequency range
the curves match the newer data sets that fall below the original
"Hamilton" data band; in the higher frequency range the curves
fall in the lower part of this band where the data for clays and
silts is located. This is precisely what we would expect since
the model predicts only intrinsic attenuation whereas the field
data in Fig. 5.13 shows the overall attenuation which includes
the effects of any scattering that has occurred. Thus we expect
a model for intrinsic attenuation to predict values near the lower
limit of the field data.

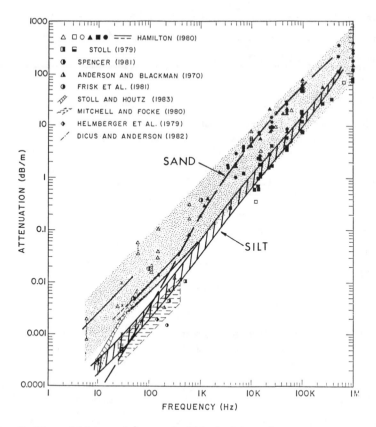

Fig. 5.13. Attenuation of dilatational waves - comparison
of model predictions with data.

SUMMARY

We have reached a point in the implementation of the Biot
theory where the response of both coarse and fine sediments can
be modeled over a wide range of frequency. The resulting variations
of frequency and, even more importantly, attenuation as a function
of frequency are markedly different from those envisioned a few
years ago based on scatter diagrams containing many high frequency
data and a few points in the low frequency range.   The theory
shows that different physical phenomena govern the transition
from low to high frequencies with overall fluid motion driven by
pore pressure gradients dominant in coarse sediments and local
fluid motion consistent with the strain rates of the skeletal
frame of increasing importance in fine materials.   The resulting

variation of attenuation coefficient, α, with frequency is significantly different from the first power dependence which has been used by many investigators to extrapolate from one frequency range to another.

CHAPTER 6
GEOACOUSTIC MODELING

FIELD EXPERIMENTS - MEASUREMENTS VS PREDICTIONS

We will now apply the formulas discussed in Chapters 1 through 5 to the prediction of velocity and attenuation as a function of frequency and depth beneath the seafloor. Examples will be given for sites where both the basic properties of the sediment (i.e., gradation, porosity, density, etc.) have been measured and, in addition, carefully made acoustic measurements (i.e., interface waves, wide angle refraction, bottom loss, etc.) have been recorded. Our objective is to estimate the acoustic response based on the known rudimentary physical parameters and then compare the results with the measured behavior.

The field data that we will discuss are largely the result of a five-year, special focus program on shallow-water acoustics sponsored by the Office of Naval Research during the period 1985 - 1990. During the summers of 1986 through 1988, a series of seismic experiments was performed in shallow water at several sites on the Atlantic coast where the geology of the near-bottom sediments was well known from prior field investigations (Stoll et al., 1988). The experiments were designed to contribute to the understanding of wave propagation in the first 100 m or so of unconsolidated sediments immediately beneath the seafloor.

The field technique used by the team from Lamont-Doherty Geological Observatory was to locate a string of gimballed geophones and an impulsive source on the seafloor in order to generate strong interface waves (i.e., Scholte - Stoneley waves) in the bottom as well as to minimize interference resulting from propagation paths in the water column. The configuration used in most of our recent experiments is shown in Fig. 6.1. After assessing the effects of wind and currents, a cable was laid on the bottom and dragged a short distance to insure uniform geophone spacing and as straight a line as possible. The ship was then anchored using a short anchor scope and the seismic source was lowered to the bottom for one or more shots.

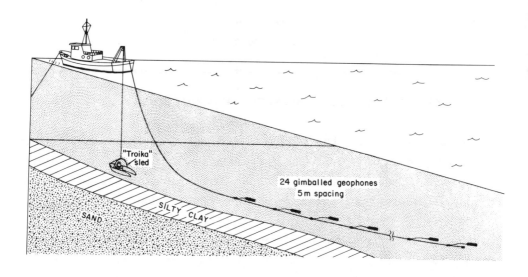

Fig. 6.1. Experimental setup for shallow water seismo-
acoustic experiments.

The source used in most of our recent work was an 8-gauge,
electrically detonated shotgun shell fired into a small "helmet"
either resting on the bottom or supported by a "Troika" sled
(i.e., a self-righting, towable sled). The blank shotgun shells
were loaded into small chambers which are sealed from the water
by a thin plastic disk which ruptures during burning. The signals
from the geophones are digitally recorded after passing through
a filter - amplifier - A/D converter system. A typical set of
raw data is shown in Fig. 6.2.

The early arrivals in Fig. 6.2 are due to compressional waves
in the water column and near-bottom sediments while the later,
low frequency arrivals are due to shear waves in the sediment and
interface waves. The compressional waves in the water and in
unlithified, near-bottom sediments propagate at speeds in the
range of 1450 to 1800 m/sec because of the dominant effect of the
fluid incompressibility whereas the shear-wave velocity in the
unlithified sediments ranges from a few meters per second to about
400 m/sec. Thus the effective Poisson's ratio in the sediment
is in the range of .45 to .49+ (see Eq. 5.5). Because of the
large differences in velocity between the dilatational waves and
the shear and interface waves, they tend to seperate at increasing
range, thus conveniently allowing for easier analysis of dispersion

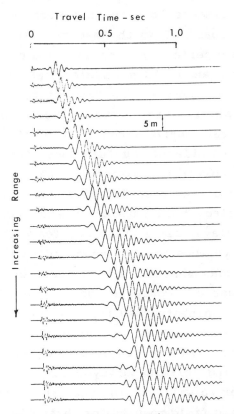

Fig. 6.2.   Raw data from a seismo-acoustic experiment in
New York Harbor over thick layer (35 m+) of sand.   Each
trace has been normalized by its maximum amplitude.

and refraction arrivals.   We have used a variety of different
techniques to analyze data such as that shown in Fig. 6.2 in order
to obtain estimates of velocity and attenuation as a function of
depth.   These include complex trace analysis (Dziewonski et al,
1969) to obtain group velocity dispersion curves, plane wave
decomposition and slant stacking (Mithal and Stoll,1989) or the
determination of cross spectra to obtain phase velocity dispersion
and a variety of techniques to analyze refraction (diving wave)
arrivals.   A detailed description of all of these techniques is
beyond the scope of this chapter so we will only include brief
descriptions of the methods used in each of the examples that are
presented.

Most of experiments to date have been performed at sites near the Atlantic coast of North America. In almost all cases the experiments were performed in areas where borings, laboratory and shipboard testing and field geophysical testing had established what has come to be known as the "ground truth". Experiments were performed at several sites studied during the USGS Atlantic Margin Coring program (AMCOR sites 6009, 6010, 6011 and 6018), at the site where the Atlantic Nuclear Generator was to have been built (AGS) and at several locations in the New York Bight and Hudson River where boring records and other data are available. The sediments sampled (both physically and acoustically) range from soft organic clayey silts which stiffen with depth as in the Hudson River to layered strata of sand and stiff clay which are found at the the AMCOR and AGS sites. One particularly interesting site is an area just off Coney Island in New York Harbor where a relatively thick, uninterrupted layer of sand is located at the bottom. This area is used as a source of sand fill for many construction projects in the New York area and so it has been heavily studied using seismic surveys, borings etc.

In this chapter we will examine the predictions of the model and the corresponding field results for three distinctly different cases: AMCOR 6011 which contains a distinct 10 m thick layer of stiff clayey sediment on the surface underlain by a relatively thick sand stratum; the site in New York Harbor containing a thick surface layer of sand; and a site in the Hudson River near the Tappan Zee Bridge containing soft surface sediments which increase in stiffness with depth.

AMCOR 6011

AMCOR site 6011 is located about 20 miles north of Atlantic City, New Jersey at $39^{\circ}43'30''$ N, $73^{\circ}58'36''$ W in water 21 m deep. The sediment strata consist of a surface layer of silty clay about 10 m thick overlying a 40 m-thick layer of sand. Beneath the sand are alternating layers of silty clay and sand of variable thickness. The unfiltered, travel-time records from one of our experiments at this site are shown in Fig. 6.2. In this figure there are several distinct arrivals that can be followed out to significant ranges. The most prominent group of wavelets is interpreted to be an interface wave moving at group velocity of about 150 m/s. The principal frequency of this arrival is about 14 or 15 Hz and there is a small amount of dispersion. Ahead of this interface wave are other groups of dispersed wavelets led by a distinct refracted SV (vertically polarized shear wave) arrival. Finally, there are higher frequency, high speed p-wave arrivals near the first break.

Geotechnical tests on samples from AMCOR 6011 (Richards, 1977) show that the upper clay layer has a voids ratio of 1.33, a Plasticity Index of 34%, a natural water content of 49%, and a specific gravity of solids of 2.71 (only one sample from this layer was analyzed). Torvane measurements gave a shear strength of 62 kPa indicating that the clay in this stratum is fairly stiff.

There were no shipboard tests to determine the physical properties of the sand layer, undoubtedly because the sand specimens are always badly disturbed and partially drained. For this reason it is necessary to assume a typical range of voids ratios that covers the possibility of both dense and loose regions existing within the sand stratum. This is not unduly restrictive since the range of voids ratios between the maximum and minimum densities of most typical uniform beach and alluvial deposits is relatively small (e.g., see Lambe and Whitman, 1970). Thus, in this example and others to follow, we have chosen a range of voids ratios from 0.5 to 0.75 in calculating overburden pressure and dynamic shear modulus in sand strata.

In the second clay stratum, which extends from a depth of about 50 m down to 80 m, a single determination of voids ratio yielded a value of 0.94, and several sets of Atterberg limits and Torvane tests give values of Plasticity Index ranging fron 30 to 40% and shear strength from 26 to 46 kPa. Thus this clay layer appears to have a somewhat higher density but lower shear strength than the surface stratum.

In order to estimate shear-wave velocity, it is necessary
to determine the mean effective stress σ_0', as a function of depth.
σ_0' may be calculated by first determining the overburden pressure
p_0 by integrating the buoyant unit weight to any depth of interest
and then using the relation $\sigma_0' = (1+2K_0)/3 \cdot p_0$, where K_0 is the
coefficient of earth pressure "at rest" in geotechnical vernacular.
The earth pressure at rest is approximated usually by a constant
within any one reasonably homogeneous stratum; however, its value
may range from values as low as 0.4 in dense sands to values near
1.0 in very soft, normally consolidated clays to values greater
than 1.0 in overconsolidated strata. Thus it is usually prudent
to use a range of values in most calculations in order to compensate
for uncertainties in the stress history, etc. In the present
example we have chosen a range of values for K_0 from .5 to 1.0.
The resulting overburden pressure and shear-wave velocity $v_s = \sqrt{\mu/\rho}$,
is shown in Fig. 6.3; μ is determined from Eq. 5.4 and ρ is the
total mass density of the sediment. A velocity determined in
this manner corresponds approximately to a "closed system" with
very small damping or to the real part of a complex velocity when
the damping is large enough to cause some measurable dispersion.

To compare the predictions shown in Fig. 6.3 the actual
response in situ, the data from one of the field experiments
performed at AMCOR site 6011 has been analyzed. A typical set
of raw data is shown in Fig. 6.4. In order to extract a velocity
model, we may analyze any one or more of the traces to obtain a
group velocity dispersion curve using complex trace analysis
(Dziewonski et al, 1969). In this method a series of closely
spaced Gaussian windows are used to obtain in-phase and quadrature
components of a signal in the frequency domain. After inverse
transformation, the amplitude of the data in the time domain is
contoured in order to obtain group velocity dispersion curves
which are located along the crests of the contour diagram. An
example of such a dispersion diagram is shown in Fig. 6.5.

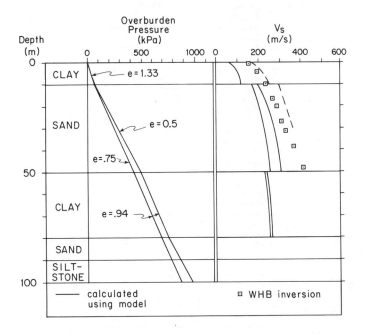

Fig. 6.3. Stratification at AMCOR 6011. Solid lines show
estimated overburden pressure and shear wave velocities.
Dashed lines show models derived from dispersion analyses.

Phase velocity dispersion curves may be obtained by calculating
the cross spectrum of any two adjacent traces. The phase angle
of the cross spectrum is then used to obtain the travel time
between the two traces for any frequency component. Because of
the possibility of spacial aliasing, some care must be exercised
in using this technique when the spacing between geophones exceeds
the wavelength at a given frequency (Bendat and Piersal, 1984).
An alternative way to determine phase velocity dispersion is by
a general wavefield transformation which utilizes the data from
all of the traces. The data is transformed to the domain of
frequency ω and horizontal ray parameter p by a plane wave
decomposition followed by a 1-D Fourier transformation of each
decomposed trace. The dispersive relationship between frequency
and horizontal phase velocity (the inverse of horizontal ray
parameter) is readily apparent in this domain. The advantage of
this wavefield transformation is that all dispersion information
in the data, including the higher modes, is recovered at the same
time.

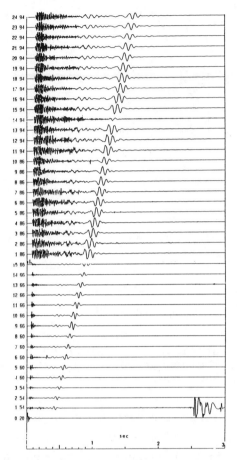

Fig. 6.4. Unfiltered travel-time curves - AMCOR 6011.
Numbers in left column identify phones and gain in dB.

Once the group and/or phase velocity dispersion curves are
determined from the field data, a range-independent model composed
of homogeneous layers of anelastic sediment is postulated, and
the phase and group velocity dispersion curves for the model are
calculated and compared with the experimental results. The model
is then modified until a reasonable correspondence between theory
and experiment is attained. The result of this kind of inverse
modeling based on the group velocity dispersion is shown by the
broken lines in Fig. 6.3. The results of a preliminary inversion
obtained by integration using the Wiechert-Herglotz-Bateman (WHB)
technique (see Stoll et al., 1988) are also shown.

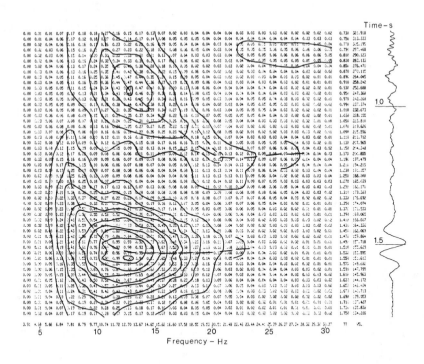

Fig. 6.5. Group velocity dispersion curve, AMCOR 6011

By comparing the model predictions based on laboratory tests shown in Fig. 6.3 and the results from field experiments, it is clear that the correct velocity depth variation is predicted but the overall amplitude is off by a significant amount. This difference should be expected for the following reasons. The modulus predicted by Eq. 5.4 reflects the first order influence of e and σ_0' in laboratory tests performed under a variety of different conditions; however, there are a number of other factors that have been shown to influence the value of the shear wave velocity observed in situ. These include the effects of aging (Humphries and Wahls, 1968), stress history (i.e., overconsolidation ratio), cyclic shear strain amplitude and sample disturbance. All of these factors tend to increase the dynamic shear modulus of an in situ sample as compared to laboratory values obtained at somewhat higher strain levels after a short period of consolidation. Various studies by geotechnical researchers have suggested that the increase will cause the value of μ to be in the range of 1.3 - 2.5 times the value obtained in the laboratory

(Arango et al, 1978, Anderson et al, 1978). Thus, the factor observed in the present case appears to be consistent with prior studies. We will make the same comparison in the other examples to be presented.

NEW YORK HARBOR

In this example we examine the results of tests carried out over a relatively thick layer of uniform sands. The test site was just offshore from Coney Island beach, New York, in an area which is used as a source of clean sands for construction projects in the New York area. Many engineering borings and seismic profiles have been made in this area (Kastens, Fray and Schubel, 1978) and they show the sand layer to be about 120 ft thick. As in the previous example, in the absence of any in-place density measurements, a range of voids ratios (.5 to .75) that will cover the normal spread expected in a deposit that contains both loose and dense regions is assumed. The resulting range of shear wave velocities based on Eq. 5.4 and a coefficient of earth pressure "at rest", K_o, of 0.5 is shown in Fig. 6.6.

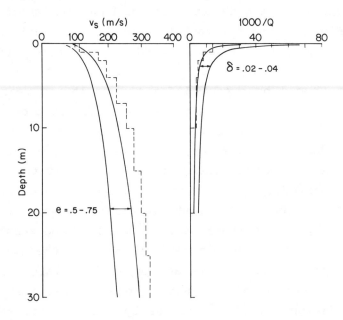

Fig. 6.6. Shear wave velocities in thick sand strata - Coney Island, New York.

A set of travel-time curves for this site is shown in Fig. 6.2 and a typical group-velocity dispersion curve in Fig. 6.7. A velocity model which produces a group velocity dispersion curve matching that of Fig. 6.7 is shown by the dashed line in Fig. 6.6 Here again the velocity profile derived from acoustic measurements in situ is the same shape as predicted by our model but displaced in such a way that a correction factor of about 1.75 is necessary.

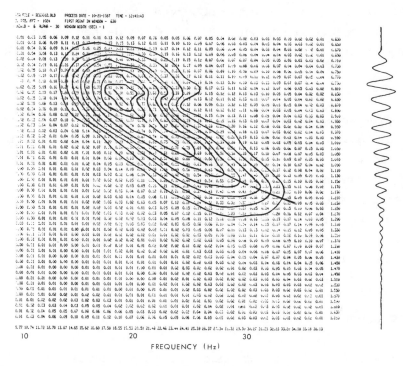

Fig. 6.7 Group velocity dispersion curve for Coney Island, New York harbor.

We have also used the data shown in Fig. 6.2 to study shear-wave attenuation as a function of depth in the sand layer. Each trace was Fourier transformed and then the variation of amplitude with range at each frequency was analyzed to determine the attenuation of the Scholte wave as a function of frequency after correction for radial spreading loss. The frequency-dependent attenuation of the Scholte wave depends on the variation of $1/Q_S$ with depth __and__ with frequency, as well as on other parameters to a lesser extent. However, since we are dealing with a data set that covers a very limited frequency range (from

10 to 30 Hz), we expect only a small change in $1/Q_S$ at each depth due to this variation in frequency whereas the the depth dependence of $1/Q_S$ is expected to be quite strong near the water-sediment interface. Hence we assume, as a first approximation, that $1/Q_S$ is independent of frequency at each depth and, using the method of partial derivatives (Takeuchi and Saito, 1972), choose a variation of $1/Q_S$ with depth that results in the proper variation of $1/Q$ with frequency for the observed Scholte wave. The results of this procedure are shown in Fig. 6.6. by dashed lines. Finally, we have estimated $1/Q_S$ as a function of depth on the basis of the formula

$$1/Q_s = K/(\sigma_0')^{.5} \qquad (6.1)$$

where K depends on shear strain amplitude. Eq. 6.1 is based on a formula given by Hardin (1965) based on very limited laboratory data. In the present case we chose K on the basis of measurements of log decrement at static stress levels approximately equal to 1 meter of embedment (e.g., see Fig. 4.8). The results of this calculation, also shown in Fig. 6.6, agree quite well with the experimental results. Both predict very rapid changes in $1/Q_S$ near the bottom.

HUDSON RIVER

The third experiment which we describe was carried out in the Hudson river at a site just north of the Tappan Zee bridge where an earlier study by Herron, Dorman and Drake was performed in 1968. In this experiment the authors chose a model based largely on the dispersion analysis of "granddaddy" waves generated by setting off explosive charges and measuring the motion at various ranges using seismographs located on the bottom. In order to make use of our model we have obtained a set of boring records made prior to the construction of the Tappan Zee bridge. The location of these borings along the bridge alignment is shown in Fig. 6.8 along with the locations of the 1968 experiment and our more recent work.

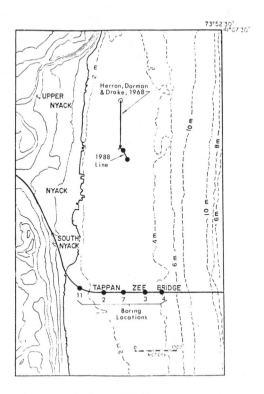

Fig. 6.8. Location of borings and experiments near
Tappan Zee Bridge, Hudson river.

The boring records show the natural moisture content of all
samples down to a depth of over 200 ft into the bottom in some
cases.   Since the natural water content is also the saturated
water content (at least where the gas content is neglegible),
voids ratio may be calculated from the formula $e = w' \cdot G_s$ where w'
is the water content expressed as a decimal and G_S is the specific
gravity of solids.   The values of e determined in this manner
from four borings are shown in Fig. 6.9.

The sediments at the bridge site consist of a surface stratum
of organic silt which varies from 10 to 30 m in thickness overlying
a thick layer of silty clay.   The organic silt is quite soft at
the water-sediment interface; however, the stiffness increases
and the water content decreases with depth in a normal manner due
to consolidation as a result of increasing overburden pressure.
An estimate of the range of overburden pressure corresponding to

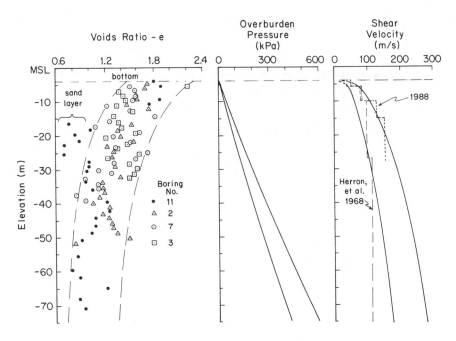

Fig. 6.9. Voids ratio, overburden pressure and shear-
wave velocity at site near Tappan Zee bridge.

the spread of data shown by the dashed lines in the first panel
is shown in the second panel of Fig. 6.9. In the third panel the
estimated shear-wave velocity based on Eqs. 5.4 and 6.1. is shown
along with a sediment model proposed by Herron et al. (1968) and
a preliminary model which we have determined on the basis of an
analysis of group velocity dispersion.

    As in the previous two examples, the velocities determined
from the field experiments, reflecting the in situ properties of
the sediment, tend to be somewhat higher than those predicted by
the sediment model which is based largely on laboratory results.
However, in all three cases (as well as for many of the cases
studied for near-surface soil deposits) it would appear that a
simple modification of Eq. 5.4 wherein the dynamic shear modulus
based on laboratory results is multiplied by a factor in the range
of 1.75 to 2.25 will bring the predicted and measured response
into acceptable agreement. Thus we modify Eq. 5.4 to read

$$\mu/p_a = a \cdot \exp(b \cdot e) \cdot (\sigma_0'/p_a)^n \cdot FF \qquad (6.2)$$

where FF is the neccesary "field factor".

USING THE MODEL AS A PREDICTIVE TOOL

At this point it will be clear to most readers that the Biot theory is not a "finished" model in that it allows a "multiple choice" determination of the 16 or so parameters suggested by Table 1.1, but rather a physically correct, general framework governing the interaction of the fluid and skeletal frame in a porous medium. The theory is consistent with all of the earlier models that have been used successfully to describe the acoustical properties of sediments (i.e., Wood's equation and Gassmann's equation which are special cases of the Biot formulation) but general enough to allow the phenominological rheology of the individual components to be varied to suit the application. While this provides us with an extremely powerful tool to study the effects of all kinds of variations in the sediment properties, it also puts a heavy burden on the user to supply meaningful input to the model. The computer-age cliche "garbage in, garbage out" certainly applies in this case. In this section we will summarize some of the empirical relations and the choice of parameters used in the examples of the previous chapters.

1. Porosity or Voids Ratio

In the examples given earlier in this chapter, it is clear that the one rudimentary variable that plays a key role in the choice of many of the other parameters is the porosity, β. In addition to controlling the total mass density, given the density of the fluid and the solids, it is important in the choice of the parameters which control the global fluid motion and the choice of the moduli of the skeletal frame. In coarse strata such as old beach deposits containing uniform sand, we expect a very limited range of porosity or voids ratio over the full range of densities that may be expected. Moreover, since in-place densities are very difficult to measure, it is recommended that a reasonable

range of voids ratio of say .5 to .75 be incorporated into the
estimates of overburden pressure, etc. and carried through the
rest of the calculations that are necessary to use the model.

In the finer marine sediments, the porosity may vary over
a wide range depending on the sediment type, the overburden
pressure and the past history of the deposit.   In areas where
direct measurements are not available, data and regression curves
relating porosity to sediment type, depth, grain size and other
variables have been given by Hamilton on numerous occasions (e.
g., see Hamilton (1976)).

2.   Density and Bulk Modulus of Grains and Pore Water

In most of our studies we have considered the variations in
density and bulk modulus of the grains and pore water to be of
secondary importance compared to the influence of other variables
such as porosity and overburden pressure.   Clearly this will not
always be the case particularly if we are interested in the effects
of water temperature or salinity, the effect of deep water or the
presence of sediment grains with unusual density or compressibility
(i. e., pelagics, volcanics, etc.). The following default values
of density and bulk modulus are built into the BASIC program given
in Appendix A; however, these are easily modified by changing
line 170 of the program:

Mass density of grains, ρ_r ------------- 2.65 g/cm^3
Mass density of pore fluid, ρ_f --------- 1.00 g/cm^3
Bulk modulus of grains, K_r ------------- 3.60x10^{11} dynes/cm^2
Bulk modulus of the fluid, K_f ---------- 2.00x10^{10} dynes/cm^2

An example of the effect of changing the compressibility of the
pore water to simulate the presence of gas was given by Stoll
(1977).

3. Variables Affecting Global Fluid Motion

Of the four variables in this category, the viscosity, η, and the "structure factor", α, are given the following default values in line 170 of the program in Appendix A:

Viscosity of pore fluid, η --------------- 1.0×10^{-2} dyne-sec/cm^2
Structure factor, α ---------------------- 1.25

The structure factor, which theoretically can vary from a value of 1.0 for parallel capillaries to 3.0 for capillaries of random orientation, was initially chosen mainly to obtain the best fit with high-frequency data of the kind found in Figs 5.4, 5.4 and 5.12. As shown by Stoll and Bryan (1970) the effect of increasing α is to decrease the attenuation in the high-frequency range. Several other investigators have also found that values slightly larger than 1 were appropriate for real particulate materials that are unlithified (i. e., see Domenico (1977) and Brunson (1983)).

The remaining two variables in this category, the permeability, k with units of (length)^2 and the "pore size parameter", a, with units of length, are the two most controversial and difficult to define for real sediments. In Biot's early papers, k represents a phenominological constant determined for the case of laminar, steady flow as in the classic Darcy experiment. In real sediments ranging from sands to clays, Lambe and Whitman (1969) have shown that this variable can range over 10 orders of magnitude. Many empirical formulas for estimating k have been proposed by different authors and there have been a number of theoretical formulations based on bundles of capillary channels or arrays of spherical particles. In almost all cases the results indicate that k depends on a size parameter related to some sort of "average or effective" size of real flow channels which in turn depends on the grain-size distribution and the porosity of the sediment. In extending the theory to consider higher frequncy motion, Biot considered the microvelocity field in simple, parallel channels (i. e.,tubes and slits), where a was a characteristic dimension   such as the diameter of the tube or the width of the slit.   Thus, it is extremely difficult to relate directly the α defined in Eq. 1.19 to the mean particle diameter and/or the porosity particularly since other variables such as grain-size distribution and particle shape also play roles in determining the actual pore geometry. The effect of grain size distribution has been studied recently by a number of investigators including Garcia-Bengochea et al. (1979) and Juang and Holtz (1986). This work has led to models for permeability that include some sort of idealized distribution of effective pore radii (i. e., ϕ-normal) which requires additional input parameters to describe.

In most of our work so far we have either used direct measurements of permeability to establish k for laboratory studies and tabulations or graphs based on experimental data, such as those given by Lambe and Whitman (1969) and many other authors, for estimating k in the field. For example, empirical equations based on experimental data from the Gulf of Mexico have been given by Bryant et al. (1975) and graphs relating permeability to the effective size at 10% finer, D_{10}, and the relative density of granular soils have been given by Burmister (1948).

In choosing values for α, especially for the coarser sediments, we have relied heavily on the trends observed in data of the kind presented in Fig. 5.4. As shown in Figs. 2.5 and 2.6, α plays the role of a scaling factor in the argument of Biot's correction factor $F(\kappa)$, and for a given frequency there is only a limited range in the values of α that will result in realistic variations of attenuation as one enters the high-frequency range. Unfortunately there is very little data in the frequency band between 200 or 300 Hz and 10 kHz because neither torsional nor pulse experiments work very well in this range. In finer grained sediments like the silt described in previous chapters, the exact value of α becomes less significant since the effects of local viscous damping become dominant.

For typical uniform beach sands, where the voids are not clogged with fines from infiltration, typical measured values of k are in the range of 10^{-6} to 10^{-8} cm^2 whereas for a finer granular material such as the silt shown by curve No. 4 in Fig. 4.2, the permeability measured in a constant head test was 2.5×10^{-10} cm^2. As was discussed in the previous chapter, even with this moderately low permeability, the effects of global fluid motion do not begin to influence the overall attenuation significantly until nearly 10 kHz is reached. In the examples shown in Fig. 5.12, k and a were 10^{-7} cm^2 and .008 cm for the sand and $2.5x10^{-10}$ cm^2 and $3.8x10^{-4}$ cm for the silt. These values of a satisfy the criterion given in chapter 5 (a = 1/6 to 1/7 of the mean grain diameter) and in the case of the sand, the attenuation curve is in good agreement with the high frequency data near the top of the Hamilton band where the sands reside.

4. Variables controlling Frequency Dependent Response of Frame

As can be seen from the first part of this chapter, variations in the shear and bulk moduli of the skeletal frame owing to rapid changes in the effective stress (i. e., overburden) play a key role in determining the variation of velocity and attenuation, particularly near the seafloor. For this reason one of the first steps in evaluating these moduli is to make the best possible estimate of overburden pressure as a function of depth using a spread of reasonable values for porosity and density of the grains. Then using Eqs. 6.1, 6.2 and 5.10, calculate $\mu(z)$ and $K(z)$. A summary of the equations and steps is given below.

a) Calculate the overburden pressure as a function of depth, z, below the seafloor using the equation

$$p_0(z) = \int_0^z \{(1-\beta)\cdot(G_s - G_w)\cdot \gamma_w\}dz$$

where the integrand is a standard expression for the buoyant unit weight of the water-saturated sediment. β is porosity, G_s and G_w are the specific gravities of the sediment grains and the water respectively, and γ_w is the unit weight of water.

b) Choose a value for the ratio of horizontal to vertical normal effective stress, $K_0 = \sigma_h{}'/\sigma_v{}'$ and use the following equation to calculate the mean effective stress

$$\sigma_0{}' = (1+2K_0)/3\cdot p_0$$

The coefficient of earth pressure "at rest", K_0, is typically in the range of .4 to .6 for sands and .4 to .7 for normally consolidated clays, depending on their Plasticity Index; however, the value may go as high as 2.5 to 3.0 in highly overconsolidated clays (Lambe and Whitman (1969)).

  c) Use Eq. 6.2 to estimate the real part of μ_1 as a function of depth below the seafloor using the following values for the constants:

$$a = 2526$$
$$b = -1.5$$
$$n = .45$$
$$FF = 2$$

  d) Estimate the imaginary part of μ_1 from the appropriate values of $1/Q$ or the logarithmic decrement determined for quasistatic conditions.

$$\mu_{1i} = 1/Q = \delta_\mu / \pi$$

From the data given in Chap. 4 we see that the low frequency asymptote of the log decrement for sands is in the range of .01 to .05 whereas for the silt tested, it falls in the range of .075 to .10.   Many of the sedimentary rocks exhibit logarithmic decrements in the range of .01 to .02. All of the values listed here assume a moist environment and reflect values measured either at a constant confining pressure, σ_0' or over a very narrow range of pressures. As a first approximation, the depth dependence of $1/Q$ or δ_μ may be incorporated using Eq. 6.1 by solving for K using a value of σ_0' corresponding to the conditions in the experiment.

  e) Choose an overall value for $\mu = \mu_1 + \mu_2$ depending on whether the sediment is coarse grained with high permeability or fine grained with lower permeability.   For coarse sediments, where global fluid motion is dominant, let $\mu = \mu_1$ determined in steps (c) and (d) above. For the finer sediments, determine μ_2 using Eq. 5.1 with $\mu_{2r} = N_r$, $\mu_{2i} = N_i$ and $N_0 = 0$. The following values for the remaining parameters in Eq. 5.1 were found to give a reasonable fit to the experimental results obtained for the silt described in previous chapters.

$\alpha = 0.2$

$\tau_0 = 1/(2\pi\{freq \ \ at \ \ peak \ \ damping\}) = 1/(2\pi \cdot \{10,000\})$

$N_\infty = \mu_{1r} \cdot \{amplitude \ \ factor\} = \mu_{1r} \cdot \{.16\}$

The evaluation of μ_2 using Eq. 5.1 has been incorporated into the BASIC program in Appendix A as an optional subroutine.

    f)   Choose the bulk modulus of the skeletal frame $K_b = K_{br} + iK_{bi}$ consistent with the shear modulus already chosen. Experiments involving p- and s-wave propagation suggest that a constant value of Poisson's ratio in the range 0.1 to 0.2 is often reasonable when considering the relation between K_b and μ_b as given by Eq. 5.6 Finally, very limited experimental results (see Fig. 4.18) suggest that the frequency dependence of 1/Q or δ is similar for both Young's modulus E_b and the shear modulus $\bar{\mu}$ so that if we choose a ratio of δ_E/δ_μ in the range of 1 to 1.5 there is sufficient information to find K_{bi} using the following relationship between the components of the complex moduli (see White, 1965, for a table of similar relationships between complex moduli).

$$K_i = \frac{3\mu_r^2 E_i - E_r^2 \mu_i}{3(3\mu_r - E_r)^2}$$

    The foregoing prescription is an example of <u>one</u> <u>of</u> <u>many</u> possible strategies for choosing input for the model. Clearly the procedure described above is slanted towards studies of the top 50 to 100 meters of sediment below the seafloor and is based largely on data derived from our own experimental work. For modeling of deeper strata, where lithification becomes important, Eq. 6.2 will no longer be valid. Cementation and other changes occurring during lithification will tend to make the complex moduli less dependent on overburden pressure and in most cases the damping will be dominated by friction and local viscous losses rather than permeability and global fluid motion. Nevertheless, the same basic framework established by Biot is still valid, with different data being used to choose the complex moduli. Recent work by various researchers in the area of rock physics such as Winkler (1985, 1988), Murphy (1982, 1984), and Bourbie (1987) are helpful in choosing meaningful input for these cases.

BIBLIOGRAPHY

Amant, W. S., "Sound Propagation in Gross Mixtures," J. Acoust. Soc. Am., 25, 638-641, 1953.

Anderson, D. G., C. Espana, and V. R. McLamore, "Estimating In Situ Shear Velocity and Moduli at Competent Sites," Proc. of the Specialty Conf. on Earthquake Engineering and Soil Dynamics, ASCE, Pasadena, ASCE, New York, 1978, 181-197.

Anderson, D. G., and F. E. Richart, Jr., "Effects of Straining on Shear Modulus of Clays," J. GED, ASCE, 102, 975-987, 1976.

Anderson, D. G., and R. D. Woods, "Comparison of Field and Laboratory Shear Moduli," in Insitu Measurement of Soil Properties, ASCE, Raleigh, North Carolina, ASCE, New York, 1975, pp. 69-92.

Anderson, R. S., and A. Blackman, "Attenuation of Low Frequency Sound Waves in Sediments," J. Acoust. Soc. Am., 49, 786-791, 1971.

Arango, I., Y. Moriwoki and F. Brown, "In-Situ and Laboratory Shear Velocity and Modulus," Proc. of the Specialty Conf. on Earthquake Engineering and Soil Dynamics, ASCE, Pasadena, ASCE, New York, 1978, 198-212.

Badiey, M., and T. Yamamoto, "Propagation of Acoustic Normal Modes in a Homogeneous Ocean Overlying Layered Anisotropic Porous Beds," J. Acoust. Soc. Am., 77, 954, 1985.

Baker, S. R., "Sound Propagation in a Superfluid Helium Filled Porous Solid: Theory and Experiment," Tech. Report No. 45, Dept. of Physics, Univ. Calif., Los Angeles, 1986.

Batchelor, G. K., An Introduction to Fluid Dynamics, Cambridge U.P., Cambridge, 1970, pp. 228.

Bazant, Z. P., and R. J. Krizek, "Saturated Sand as an Inelastic Two-Phase Medium," J. Eng. Mech. Div., ASCE, 101, 317-332, 1975.

Bedford, A., and R. D. Costley and M. Stern, "On the Drag and Virtual Mass Coefficients in Biot's Equations," J. Acoust. Soc. Am., 76, 1804, 1984.

Biot, M. A., "General Theory of Three-Dimensional Consolidation," J. Applied Phys., 12, 155-164, 1941.

Biot, M. A., "The Interaction of Rayleigh and Stoneley Waves in the Ocean Bottom," Bull. Seism. Soc. Am., 42, 81-93, 1952.

Biot, M. A., "Theory of Elastic Waves in a Fluid-Saturated Porous Solid. I. Low Frequency Range," J. Acoust. Soc. Am., 28, 168-178, 1956a.

Biot, M. A., "Theory of Elastic Waves in a Fluid-Saturated Porous Solid. II. Higher Frequency Range," J. Acoust. Soc. Am., 28, 179-191, 1956b.

Biot, M. A., "Mechanics of Deformation and Acoustic Propagation in Porous Dissipative Media," J. Appl. Phys., 33, 1482-1498, 1962a.

Biot, M. A., "Generalized Theory of Acoustic Propagation in Porous Dissipative Media," J. Acoust. Soc. Am., <u>34</u>, 1254-1264, 1962b.

Biot, M. A., "Non-linear and Semi-linear Rheology of Porous Solids," J. Geophys. Res., <u>78</u>, 4924-4937, 1973.

Biot, M. A., and D. G. Willis, "The Elastic Coefficients of the Theory of Consolidation," J. Appl. Mech., <u>24</u>, 594-601, 1957.

Bishop, A. W., and D. J. Henkel, <u>The Measurement of Soil Properties in the Triaxial Test</u>, Edward Arnold, Ltd., London, 1957.

Blevins, R., <u>Formulas for Natural Frequency and Mode Shape</u>, Van Nostrand-Reinhold, New York, 1979, pp 107.

Bocherdt, R. D., "Energy and Plane Waves in Linear Viscoelastic Media," J. Geophs. Res., <u>78</u>, 2442-2453, 1973.

Bocherdt, R. D., "Reflection and Refraction of Type-II S Waves in Elastic and Anelastic Media," Bull. Seis. Soc. Am., <u>67</u>, 43-67, 1977.

Bourbie, T., O. Coussy, and B. Zinsner, <u>Acoustics of Porous Media</u>, Gulf, Houston, 1987.

Brandt, H., "A Study of the Speed of Sound in Porous Granular Media," J. Appl. Mech. <u>22</u>, 479-486, 1955.

Brandt, H., "Factors Affecting Compressional Wave Velocity in Unconsolidated Marine Sand Sediments," J. Acoust. Soc. Am. <u>32</u>, 171-179, 1960.

Brekhovskikh, L. M., <u>Waves in Layered Media</u>, Academic Press, New York, 1960, pp. 15-22.

Brunson, B. A., "Shear Wave Attenuation in Unconsolidated Laboratory Sediments," thesis for Ph.D. degree, Oregon State University, School of Oceanography, 1983.

Brunson, B. A. and R. K. Johnson, "Laboratory Measurements of Shear-Wave Attenuation in Saturated Sands," J. Acoust. Soc. Am. <u>68</u>, 1371-1375, 1980.

Bryan, G. M., "A Compaction Model for Compressional Wave Velocity," J. Acoust. Soc. Am. <u>76</u>, 192-197, 1984.

Bryan, G. M. and R. D. Stoll, "The Dynamic Shear Modulus of Marine Sediments," J. Acoust. Soc. Am. <u>83</u>, 2159-2164, 1988.

Bryant, W. R., W. Hottman, and P. Trabant, "Permeability of Unconsolidated and Consolidated Marine Sediments, Gulf of Mexico," Marine Geotechnology, <u>1</u>, 1-14, 1975.

Buchen, P. W., "Plane Waves in Viscoelastic Media," Geophys. J. R. Soc. <u>23</u>, 531-542, 1971.

Bucker, H. P., J. A. Whitney, and K. L. Keir, "Use of Stonely Waves to Determine the Shear Velocity in Ocean Sediments," J. Acoust. Soc. Am. <u>36</u>, 1595-1596, 1964.

Burmister, D. M., "The Importance and Practical Use of Relative Density in Soil Mechanics," Proc. Am. Soc. Testing and Mat., <u>48q</u>, 1948.

Burmister, D. M., "Physical, Stress-Strain, and Strength Responses of Granular Soils," Spec, Tech. Publ. no. 322, 67-97, 1962.

Busby, J., and E. G. Richardson, "The Absorption of Sound in Sediments," Geophysics <u>22</u>, 821-828, 1957.

Carman, P. C., *Flow of Gases Through Porous Media*, Academic Press, New York, 1956.

Cattaneo, C., "Sul Contatto di due corpi elastici," Academia dei Lincei, Rendiconti 27, 342-348, 434-436, and 474-478, 1938.

Christensen, R. E., J. A. Frank, and W. H. Geddes, "Low Frequency Propagation Via Shallow Refracted Paths Through Deep Ocean Unconsolidated Sediments," J. Acoust. Soc. Am. 57, 1421-1426, 1975.

Clay, C. S., and H. Medwin, *Acoustical Oceanography*, John Wiley and Sons, New York, 1977, pp. 487-488.

Cole, K. S. and R. H. Cole, "Dispersion and Adsorption in Dielectrics," J. Chem. Physics 9, 341-351, 1941.

Cooper, H. F. Jr., "Reflection and Transmission of Oblique Plane Waves at a Plane Interface between Viscoelastic Media," J. Acoust. Soc. Am. 42, 1064-1069, 1967.

Cooper, H. F. Jr., and E. L. Reiss, "Reflection of Plane Viscoelastic Waves from Plane Boundaries," J. Acoust. Soc. Am. 39, 1133 -1138, 1966.

Deresiewicz, H., "Stress-Strain Relations for a Simple Model of a Granular Medium," J. Appl. Mech. 25, 402-406, 1958.

Deresiewicz, H., and J. T. Rice, "The Effect of Boundaries on Wave Propagation in a Liquid-Filled Porous Solid III. Reflection of Plane Waves at a Free Boundary (General Case)," Bull. S.S.A. 54, 595-625, 1962.

Deresiewicz, H., and J. T. Rice, "The Effect of Boundaries on Wave Propagation in a Liquid-Filled Porous Solid V. Transmission Across a Plane Interface," Bull S.S.A. 54, 409-416, 1964.

Deresiewicz, H., and R. Skalak, "On Uniqueness in Dynamic Poroelasticity," Bull. S.S.A. 53, 783-788, 1963.

Dicus, R. L., and R. S. Anderson, "Effective Low Frequency Geoacoustic Properties Inferred from Measurements in the Northeast Atlantic," Naval Ocean Research and Development Activity Report No. 21, 1982.

Digby, P. J., "The Effective Moduli of Porous Granular Rocks," J. Appl. Mech. 48, 803, 1981.

Domenico, S. N., "Effect of Water Saturation on Seismic Reflectivity of Sand Reservoirs Encased in Shale," Geophysics 39, 759-769, 1974.

Domenico, S. N., "Elastic Properties of Unconsolidated Porous Sand Reservoirs," Geophysics, 42, 1339-1368, 1977.

Drnevich, V. P., "Effects of Strain History on the Dynamic Properties of Sand," Ph.D. dissertation, University of Michagan, 1967.

Drnevich, V. P., and F. E. Richart, Jr., "Dynamic Prestraining of Dry Sand," J. Soil Mech. and Found. Div., ASCE 96, 453-469, 1970.

Duffy, J., "A Differential Stress-Strain Relation for the Hexagonal Close-Packed Array of Elastic Spheres," J. Appl. Mech. 26, 88-94, 1959.

Duffy, J., and R. D. Mindlin, "Stress-Strain Relations and Vibrations of a Granular Medium," J. Appl. Mech. 79, 585-593, 1957.

Dunn, K. J. "Acoustic Attenuation in Fluid Saturated Porous Cylinders at Low Frequencies," J. Acoust. Soc. Am. 79, 1709-1721, 1986.

Dunn, K. J. "Sample Boundary Effects in Acoustic Attenuation of Fluid Saturated Porous Cylinders," J. Acoust. Soc. Am. 81, 1259-1266, 1987.

Dziewonski, A., S. Block and M. Landisman, "A Technique for Analysis of Transient Seismic Signals," Bull. Seismol. Soc. Am. 59, 427-444, 1969.

Frisk, G. V., J. A. Doutt and E. E. Hays, "Bottom Interaction of Low Frequency Acoustic Signals at Small Grazing Angles in the Deep Ocean," J. Acoust. Soc. Am. 69, 84-89, 1981.

Frisk, G. V., J. A. Doutt and E. E. Hays, "Geoacoustic Models of the Icelandic Basin," J. Acoust. Soc. Am. 80, 84-89, 1986.

Garcia-Bengochea, I., C. W. Lovell, and A. G. Altshaeffl, "Relation between Pore Size Distribution and Permeability of Silty Clay," J. Geotechn. Engr., Am. Soc. Civil Engr., 105, 839-859, 1979.

Gardner, G. H. F., M. R. J. Wyllie, and D. M. Droschak, "The Effects of Pressure and Fluid Saturation on the Attenuation of Elastic Waves in Sands," J. Petr. Tech., Feb., 189-198, 1964.

Gassmann, F., "Uber die Elastizitat Poroser Medien," Bierteljahrschr. Naturforsch. Ges., Zurich, 96, 1-23, 1951.

Geertsma, J., "Velocity-Log Interpretation: The Effect of Rock Bulk Compressibility," J. Soc. Petrol. Engrs., Dec., 235-248, 1961.

Geertsma, J., and D. C. Smit, "Some Aspects of Elastic Wave Propagation in Fluid-Saturated Porous Solids," Geophysics 26, 169-181, 1957.

Hall, J. R., Jr., and F. E. Richard, Jr., "Dissipation of Elastic Wave Energy in Granular Solid," J. Soil Mech. Found. Div., A.S.C.E. 89(SM6), 27-56, 1963.

Hamilton, E. L., "Elastic Properties of Marine Sediments," J. Geophys. Res. 76, 579-604, 1971.

Hamilton, E. L., "Compressional Wave Attenuation in Marine Sediments," Geophysics 37, 620-646, 1972.

Hamilton, E. L., "Geoacoustic Models of the Sea Floor," in Physics of Sound in Marine Sediments edited by L. Hampton (Plenum Press, New York, 1974).

Hamilton, E. L., "Attenuation of Shear Waves in Marine Sediments," J. Acoust. Soc. Am. 60, 334-338, 1976a.

Hamilton, E. L., "Shear Wave Velocity Versus Depth in Marine Sediments: A Review," Geophysics 41, 985-986, 1976b.

Hamilton, E. L., "Sound Attenuation as a Function of Depth in the Sea," J. Acoust. Soc. Am. 59, 528-535, 1976c.

Hamilton, E. L., "Variations of Density and Porosity with Depth in Deep-Sea Sediments," J. Sed. Petrol. <u>46</u>, 280-300, 1976d.

Hamilton, E. L., "Geoacoustic Modeling of the Sea Floor," J. Acoust. Soc. Am. <u>68</u>, 1313-1340, 1980.

Hamilton, E. L., "Acoustic Properties of Sediments," in <u>Acoustics and the Ocean Bottom</u>, edited by A. Lara-Saenz, C. Ranz Guerra, and C. Carbo-Fite (Consejo Superior de Invesigaciones Cientificas [CSIC], Madrid, 1987).

Hampton, L. D., "Acoustic Properties of Sediments," J. Acoust. Soc. Am. <u>42</u>, 882-890, 1967.

Hampton, L. D., and A. L. Anderson, "Acoustics and Gas in Sediments - Applied Research Laboratories Experience," in <u>Natural Gases in Marine Sediments</u>, edited by I. R. Kaplan, Plenum, New York, 1974, pp. 249-273.

Hanna, J. S., "Short-Range Transmission Loss and the Evidence for Bottom Refracted Energy," J. Acoust. Soc. Am. <u>53</u>, 1686-1690, 1973.

Hardin, B. O., "Dynamic Versus Static Shear Modulus for Dry Sands," Mater. Res. and Stand., ASTM, <u>5</u>, 232-235, 1965a.

Hardin, B. O., "The Nature of Damping in Sands," J. Soil Mech. Found. Engr. Div., ASCE <u>91</u>, SMI, 63-97, 1965b.

Hardin, B. O., "The Nature of Stress-Strain Behavior for Soils," in Proc. for the Specialty Conf. on Earthquake Engineering and Soil Dynamics, ASCE, New York, 1978, pp.2-90.

Hardin, B. O., and W. L. Black, "Vibration Modulus of Normally Consolidated Clay," J. Soil Mech. Found. Div., ASCE <u>94,</u> SM2, 353-369, 1968.

Hardin, B. O., and U. P. Drnevich, "Shear Modulus and Damping in Soils: Measurement and Parameter Effects," J. Soil Mech. Found. Div., ASCE <u>98</u>, SM6, 603-624, 1972.

Hardin, B. O., and J. Music, "Apparatus for Vibration of Soil Specimens During the Triaxial Test," in <u>Instruments and Apparatus for Soil and Rock Mechanics</u>, ASTM Spec. Tech Publ. 392 (Am. Soc. Testing Materials, Philadelphia, 1965) pp. 55-72.

Hardin, B. O., and F. E. Richart, Jr., "Elastic Wave Velocities in Granular Soils," J. Soil Mech. Found. Div., ASCE, <u>89</u>, SM1, 33-65, 1963.

Helmberger, D. V., G. Engen, and P. Scott, "A Note on Velocity, Density and Attenuation Models for Marine Sediments Determined from Multibounce Phases," J. Geophys. Res. <u>84</u>, 667-671, 1979.

Hendron, A. J., "The Behavior of Sand in One-Dimensional Compression," thesis presented in partial fulfillment of the requirements for the degree of Doctor of Philosophy, Univ. of Ill., 1963.

Hovem, J. M. and G. D. Ingram, "Viscous Attenuation of Sound in Saturated Sand," J. Acoust. Soc. Am. <u>66</u>, 1807-1817, 1979.

Hovem, J. M. "Viscous Attenuation of Sound in Suspensions and High Porosity Marine Sediments," J. Acoust. Soc. Am. <u>67</u>, 1559-1563, 1980.

Humphries, W. K., and H. E. Wahls, "Stress History Effects on the Dynamic Modulus of Clay," J. Soil Mech. and Found. Div., ASCE, 94, 371-389, 1968.

Iwasaki, T., F. Tatsuoka and Y. Takagi, "Shear Modulus of Sands Under Cyclic Torsional Shear Loading," Soils and Foundations, 1977.

Jackson, K. C., Textbook of Lithology, McGraw-Hill Book Co., New York, 1970, p. 97.

Jensen, F. B., and H. Schmidt, "Shear Properties of Ocean Sediments determined from Numerical Modeling of Scholte Wave Data," in Ocean Seismo-Acoustics, edited by T. Akal and J. M. Berkson, Plenum, New York, 1986, pp. 683-692.

Johnson, D. L., and T. J. Plona, "Acoustic Slow Waves and the Consolidation Transition," J. Acoust. Soc. Am. 72, 556-565, 1982.

Johnson, D. L.,J. Koplik and R. Dashen, "Theory of Dynamic Permeability and Tortuosity in Fluid-Saturated Porous Media," J. Fluid Mech., 176, 379, 1987.

Juang, C. H., and R. D. Holtz, "Fabric, Pore-size Distribution and Permeability of Sandy Soils," J. Geotechn. Engrg., Am. Soc. Civil Engr., 112, 855-868, 1986.

Kastens, K. A., C. T. Fray and J. R. Schubel, "Environmental Effects of Sand Mining in the Lower Bay of New York Harbor," Spec. Report No. 15, Marine Sciences Research Center, State Univ. of N. Y., 1978.

Kibblewhite, A. C., "Attenuation of Sound in Marine Sediment: A Review with Emphasis on New Low Frequency Data," Applied Research Laboratory Report ARL-TP-88-10, 1988.

Koutsoftos, D. C., and J. A., Fischer, "Dynamic Properties of Two Marine Clays," J. Geotech. Div., ASCE 106, 645-657, 1980.

Kryloff, N., and N. Bogoliuboff, Introduction to Non-Linear Mechanics, Princeton U., Princeton, N.J., 1947, pp.55-63.

Lambe, T. W., and R. V. Whitman, Soil Mechanics, John Wiley and Sons, Inc., New York, 1969, pp. 281-294.

Lawrence, F. V., Jr., "Ultrasonic Shear Wave Velocities in Sand Clay," Res. Rept. R65-05, Soils Pub. No. 175, Dept. of Civil Eng., M.I.T., 1965.

Laughton, A. S., "Laboratory Measurements of Seismic Velocities in Ocean Sediments," Proc. Roy. Soc. (London), Ser. A, 222, 336-341, 1954.

Laughton, A. S., "Sound Propagation in Compacted Ocean Sediments," Geophysics 22, 233-260, 1957.

Lee, T-M, "Method of Determining Properties of Viscoelastic Solids Employing Forced Vibrations," J. Appl. Physics 34, 1524-1529, 1963

Mackenzie, K. V., "Reflection of Sound from Coastal Bottoms," J. Acoust. Soc. Am. 32, 221-231, 1960.

McCann, C., and D. M. McCann, "The Attenuation of Compressional Waves in Marine Sediments," Geophysics 34, 882-892, 1969.

McLeroy, E. C., and A. DeLoach, "Sound, Speed and Attenuation, from 15 to 1500 kHz, Measured in Natural Sea-Floor Sediments," J. Acoust. Soc. Am. 44, 1148-1150, 1968.

Menke, W., "A Formula for the Apparent Attenuation of Acoustic Waves in Randomly Layered Media," Geophys. J. R. Astr. Soc. 75, 541-544, 1983.

Mindlin, R. D., "Compliance of Elastic Bodies in Contact," J. Appl. Mech. 16, 259-268, 1949.

Mindlin, R. D., "Mechanics of Granular Media," Proc. 2nd U.S. Nat. Cong. Appl. Mech, ASME, New York, 1954, pp. 13-20.

Mindlin, R. D., and H. Deresiewicz, "Elastic Spheres in Contact Under Varying Oblique Forces," J. Appl. Mech. 20, 327-344, 1953.

Mitchell, S. K., and K. C. Focke, "New Measurements of Compressional Wave Attenuation," J. Acoust. Soc. Am. 67, 1582-1589, 1980.

Mitchell, S. K., and K. C. Focke, "The Role of the Seabed Attenuation Profile in Shallow Water Acoustic Propagation," J. Acoust. Soc. Am. 73, 466-473, 1983.

Mithal, R. and R. D. Stoll, "Wavefield Transformation to Recover Dispersion and Determine S-Wave Velocity in the Uppermost Sediments," (in preparation), 1989.

Murphy, W. F., III, "Effects of Partial Water Saturation on Attenuation in Massillon Sandstone and Vycor Porous Glass," J. Acoust. Soc. Am. 71, 1458-1468, 1982.

Murphy, W. F., III, "Acoustic measures of Partial Gas Saturation in Tight Sandstones," J. Geophys. Res. 89, 11549-11559, 1984.

Murphy, W.F., III, K.W. Winkler and R.L. Kleinberg, "Contact Microphysics and Viscous Relaxation in Sandstone," in Physics and Chemistry of Porous Media edited by D.L. Johnson and P.N. Sen, Am. Inst. Phys., New York, 1984, pp. 176-190.

Nolle, A.W., W.A. Hoyer, J.F. Misfud, W.R. Runyan, and M.B. Ward, "Acoustical Properties of Water-Filled Sands," J. Acoust. Soc. Am., 35, 1394-1408, 1963.

Ogushwitz, P. R., "Applicability of the Biot Theory, I. Low-Porosity Materials; II. Suspensions; III. Waves Speeds Versus Depth in Marine Sediments," J. Acoust. Soc. Am. 77, 429, 1985.

Plona, T. J., "Observation of a Second Bulk Compressional Wave in a Porous Medium at Ultrasonic Frequencies," Appl. Phys. Lett. 36, 259-261, 1980.

Pollard, H. F., Sound Waves in Solids, Pion Limited, London, 1977.

Rasolofosaon, P., "Propagation Des Ondes Acoustiques Dans Les Milieux Poreux - Effets D'Interface - (Theory et Experiences)'" Thesis presented to the Univ. of Paris for the degree of Docteur es Sciences Physiques, 1987.

Rasolofosaon, P., "Importance of Interface Hydraulic Condition on the Generation of Second Bulk Compressional Wave in Porous Media," Appl. Phys. Lett. 52, 780-782, 1988.

Richards, P. G., and W. Menke, "The Apparent Attenuation of a Scattering Medium," Bull. Seismol. Soc. Am. 73, 1005-1021, 1983.

Richart, F. E., Jr., J. R. Hall, Jr., and R. D. Woods, Vibration of Soils and Foundations, Prentice-Hall, Englewood Cliffs, 1970.

Schoenberg, M., "Transmissions and Reflection of Plane Waves at an Elastic-Viscoelastic Interface," Geophys. J. R. Soc. 25, 35-47, 1971.

Shumway, G., "Sound Speed and Absorption Studies of Marine Sediments by a Resonance Method," Geophysics 25, 451-467, 659-682, 1960.

Spencer, J. W., "Stress Relaxations at Low Frequencies in Fluid Saturated Rocks: Attenuation and Modulus Dispersion," J. Geophys. Res. 86, 1803-1812, 1981, (c) American Geophysical Union.

Spinner, S. and W. E. Teft, "A Method for Determining Mechanical Resonance Frequencies and for Calculating Elastic Moduli from these Frequencies," Proc. A.S.T.M., 61, 1221-1238, 1961.

Stern, M., A. Bedford and H.R. Millwater, "Wave Reflection from a Sediment Layer with Depth-Dependent Properties," J. Acoust. Soc. Am. 77, 1781, 1985.

Stokey, W. F., "Vibration of Systems having Distributed Mass and Elasticity," in Shock and Vibration Handbook, edited by C. Harris and C. Crede, 2nd Ed., McGraw-Hill, 1976.

Stokoe, K.H., S.H. Lee and D.P. Knox, "Shear Moduli Measurements under True Triaxial Stresses," in Advances in the Art of Testing Soils under Cyclic Conditions edited by V. Khosla, ASCE, New York, 1985, pp. 166-185.

Stoll, R. D., "Essential Factors in a Mathematical Model of Granular Soil," Proc. Internat. Symp. on Wave Propagation and Dynamic Properties of Earth Materials, Univ. New Mexico Press, Albuquerque, 1968, pp. 201-209.

Stoll, R. D., "Acoustic Waves in Saturated Sediments," in Physics of Sound in Marine Sediments edited by L. Hampton, Plenum, New York, 1974, pp. 19-39.

Stoll, R. D., "Acoustic Waves in Ocean Sediments," Geophysics 42, 715-725, 1977.

Stoll, R. D., "Damping in Saturated Soil," in Proc. Specialty Conf. on Earthquake Engrg. and Soil Dynamics, ASCE, New York, 1978, pp. 960-975.

Stoll, R. D., "Experimental Studies of Attenuation in Sediments," J. Acoust. Soc. Am. 66, 1152-1160, 1979.

Stoll, R. D., "Theoretical Aspects of Sound Transmission in Sediments," J. Acoust. Soc. Am. 68, 1341-1350, 1980.

Stoll, R. D., "Computer-Aided Measurements of Damping in Marine Sediments," in Computational Methods and Experimental Measurements edited by G. A. Keramidas and C. A. Brebbia, ISCME, Computational Mechanics Publications, Southampton, England, 3.29-3.39, 1984.

Stoll, R. D., "Computer-Aided Studies of Complex Soil Moduli," in Measurement and Use of Shear Wave Velocity for Evaluating Dynamic Soil Properties, ed. R. Woods, (ASCE, New York, 1985a) pp. 18-33.

Stoll, R. D., "Marine Sediment Acoustics," J. Acoust. Soc. Am. 77, 1789-1799, 1985.

Stoll, R. D., "Acoustic Waves in Marine Sediments," in Ocean Seismo-Acoustics edited by T. Akal and M. Berkson, Plenum, New York, 1986, pp. 417-434.

Stoll, R. D., "Stress-Induced Anisotropy in Sediment Acoustics," J. Acoust. Soc. Am. 85, 702-708, 1989.

Stoll, R. D., "Geoacoustic Properties of a Marine Silt," in Clay Microstructure: From Mud to Shale edited by R. Bennett, NORDA, Springer-Verlag, New York, 1990.

Stoll, R. D., G. M. Bryan, R. Flood, D. Chayes, and P. Manley, "Shallow Seismic Experiments Using Shear Waves," J. Acoust. Soc. Am., 83, 93-102, 1988.

Stoll, R. D., and G. M. Bryan, "Wave Attenuation in Saturated Sediments," J. Acoust. Soc. Am. 47, 1440-1447, 1970.

Stoll, R. D., and I. A. Ebeido, "Unloading Effects in the Dynamic Response of Granular Soil," Exper. Mech. 6, 1966.

Stoll, R. D., and R. E. Houtz, "Attenuation Measurements with Sonobuoys," J. Acoust. Soc. Am. 73, 163-172, 1983.

Stoll, R. D., and T. K. Kan, "Reflection of Acoustic Waves at a Water-Sediment Interface," J. Acoust. Soc. Am. 70, 149-156, 1981.

Thurston, C. W. and H. Deresiewicz, "Analysis of a Compression Test of a Model of a Granular Medium," J. Appl. Mech. 26, 251-258, 1959.

Timoshenko, S., and J. N. Goodier, Theory of Elasticity, McGraw-Hill, New York, 1951, p. 372.

Tittmann, B. R., J. R. Bulan and M. Abdel-Gawad, "Dissipation of Elastic Waves in Fluid Saturated Rocks," in Physics and Chemistry of Porous Media edited by D.L. Johnson and P.N. Sen, Am. Inst. Phys. New York, 1984, pp. 131-143.

Whitman, R. V., E. T. Miller and P. J. Moore, "Yielding and Locking of Confined Sand," J. Soil Mech. and Found. Engrg. ASCE 90, SM4, 57-84, 1964.

White, J. E., Seismic Waves - Radiation, Transmission, and Attenuation (McGraw-Hill, New York, 1965).

Winkler, K. W., "Dispersion of Velocity and Attenuation in Berea Sandstone," J. Geophys. Res. 90, 6793-6800, 1985.

Winkler, K. W., "Frequency Dependent Ultrasonic Properties of High Porosity Sandstones," J. Geophys. Res. 88, 9493-9499, 1988.

Wood, A. B., A Textbook of Sound, G. Bell and Sons, London, 1911.

Wood, A. B., and D. E. Weston, "The Propagation of Sound in Mud," Acoustica, 14, 156-162, 1964.

Yamamoto, T., "Acoustic Propagation in the Ocean with a Poro-Elastic Bottom," J. Acoust. Soc. Am. 73, 1587, 1983.

Yamamoto, T. and A. Turgut, "Acoustic Wave Propagation through Porous Media with Arbitrary Pore Size Distribution," J. Acoust. Soc. Am. 83, 1744-1751, 1988

BASIC PROGRAM TO CALCULATE VELOCITY, ATTENUATION AND LOG DECREMENT USING THE BIOT MODEL

```
1 '************ BIOT - STOLL MODEL ****************
2 'This program calculates the velocity and attenuation of waves according to
    the Biot theory.
3 'Three mechanisms for energy dissipation are included:
4 '(1)  Intergranular friction as determined by the low frequency asymptote of
          the log decrement assigned to the frame moduli.
5 '(2)  Local viscous losses determined by the "Cole-Cole" viscoelastic model
          which controls the frequency dependent reponse of the skeletal frame.
6 '(3)  Global viscous losses as determined by the viscodynamic operators in
          the basic Biot formulation.
7 'The program contains the option to include only friction and global viscous
    losses (i.e., as in the case of coarse sediments of high permeability).

20 WIDTH "LPT1:",132
30 DEFINT I,J,N
40 DIM F(20),IP(4),DEC(20)
50 FF$="##.###^^^^  ":FFF$="#####.##  ":FFFF$="##.#### ":FZ$="##.##^^^^"
60 PI=4*ATN(1):R=SQR(2)
70 INPUT "HEADING";H$
80 INPUT "POROSITY";PHI
90 INPUT "PERMEABILITY (CM**2)";PERM
100 INPUT "RADIUS OF AVER. PORE (CM)";RAD
110 INPUT "SHEAR MODULUS OF FRAME (DYNES/CM**2)";GR
120 INPUT "POISSONS RATIO OF FRAME";FRMNU
125 PRINT "choice of variable complex modulus includes effects of local
    viscous losses in model"
130 INPUT "CONSTANT OR VARIABLE COMPLEX MODULUS (C OR V)";G$
140 IF G$="C" OR G$="c" THEN INPUT "SHEAR LOG DECREMENT";GDEC
145 PRINT "preliminary experimental work suggests that the ratio of log
    decrements EDEC/GDEC will be in the range of 1 to 1.5"
150 INPUT "EDEC/GDEC";EGRAT
160 READ RMOD,FMOD,RDEN,FRHO,VISCF,FAC
170 DATA 3.6E11,2.0E10,2.65,1.0,.01,1.25
175 RHO=RDEN*(1-PHI)+PHI*FRHO
180 LPRINT CHR$(15):LPRINT DATE$, TIME$:LPRINT
190 LPRINT" ":LPRINT" ":LPRINT TAB(10) "****************************":
    LPRINT TAB(10) H$:LPRINT TAB(10) "****************************":LPRINT" "
200 LPRINT TAB(10) "SPECIFIC GRAVITY OF GRAINS ---------------------";RDEN
210 LPRINT TAB(10) "SPECIFIC GRAVITY OF FLUID ----------------------";FRHO
220 LPRINT TAB(10) "AB VISCOSITY OF FLUID (DYNE-SEC/CM**2) ---------";VISCF
230 LPRINT TAB(10) "BULK MOD OF GRAINS (DYNES/CM**2)----------------";RMOD
240 LPRINT TAB(10) "BULK MOD OF FLUID (DYNES/CM**2)-----------------";FMOD
250 LPRINT TAB(10) "DENSITY COUPLING FACTOR ------------------------";FAC
260 LPRINT TAB(10) "POROSITY --------------------------------------";PHI
270 LPRINT TAB(10) "PERMEABILITY (CM**2) --------------------------";PERM
280 LPRINT TAB(10) "RADIUS OF AVER. PORE (CM) ---------------------";RAD
300 LPRINT TAB(10) "SHEAR MOD OF FRAME (DYNES/CM**2) --------------";GR
310 LPRINT TAB(10) "POISSONS RATIO --------------------------------";FRMNU
320 LPRINT TAB(10) "RATIO EDEC/GDEC -------------------------------";EGRAT
330 LPRINT TAB(10) "TOTAL SPECIFIC GRAVITY ------------------------";RHO
340 IF G$="V" OR G$="v" THEN GOSUB 5000
350 IF G$="C" OR G$="c" THEN LPRINT TAB(10) "LOG DECREMENT FOR SHEAR
```

```
                  ----------------------";GDEC
360 INPUT "NO. FREQUENCIES";NFRQ
370 FOR I = 1 TO NFRQ
380 INPUT "FREQ";F(I):GOSUB 6000
390 NEXT I
480 LPRINT TAB(10) "FREQUENCY     V1              A1              DEC1
          VS              AS              DECS    V2      DEC2    GDEC"
490 LPRINT TAB(10) " (HZ)      (M/SEC)    (N/M)     (DB/M)
          (M/SEC)     (N/M)      (DB/M)                  (M/SEC)"
500 FOR I=1 TO NFRQ
510 GDEC = DEC(I):GI=GDEC*GR/PI
530 EMODR=(FRMNU+1)*GR*2
532 EMODI=EGRAT*GI/GR*EMODR
534 BR=GR*EMODR/(3*GR-EMODR)/3
536 BI=(3*GR*GR*EMODI-EMODR*EMODR*GI)/(3*GR-EMODR)/(3*GR-EMODR)/3
538 BDEC=BI/BR*PI
540 M=FAC*FRHO/PHI
541 D=RMOD*(PHI*(RMOD/FMOD-1)+1)
542 DNOM=(D-BR)*(D-BR)+BI*BI
543 HR=(((RMOD-BR)*(RMOD-BR)-BI*BI)*(D-BR)+2*BI*BI*(RMOD-BR))/DNOM+BR+4*GR/3
544 HI=(((RMOD-BR)*(RMOD-BR)-BI*BI)*BI-2*BI*(RMOD-BR)*(D-BR))/DNOM+BI+4*GI/3
545 KR=((RMOD*RMOD-BR*RMOD)*(D-BR)+BI*RMOD*BI)/DNOM
546 KI=((RMOD*RMOD-BR*RMOD)*BI-BI*RMOD*(D-BR))/DNOM
547 DR=RMOD*RMOD*(D-BR)/DNOM
548 DI=RMOD*RMOD*BI/DNOM
550 Z=SQR(F(I)*2*PI*FRHO/VISCF)*RAD
560 IF Z>8 GOTO 810
570 BER=1:DBER=0:SIGN=-1
580 FOR J=4 TO 32 STEP 4
590 DEN=1
600 FOR JJ=2 TO J STEP 2
610 DEN=JJ*JJ*DEN
620 NEXT JJ
630 BER=SIGN*Z^J/DEN+BER
640 DBER=SIGN*Z^(J-1)/DEN*J+DBER
650 SIGN=-SIGN
660 NEXT J
670 BBEI=0:DDBEI=0
680 SIGN=1
690 FOR J=2 TO 30 STEP 4
700 DEN=1
710 FOR JJ=2 TO J STEP 2
720 DEN=JJ*JJ*DEN
730 NEXT JJ
740 BBEI=SIGN*Z^J/DEN+BBEI
750 DDBEI=SIGN*Z^(J-1)/DEN*J+DDBEI
760 SIGN=-SIGN
770 NEXT J
780 TR=(BBEI*DDBEI+BER*DBER)/(BER*BER+BBEI*BBEI)
790 TI=(BER*DDBEI-BBEI*DBER)/(BER*BER+BBEI*BBEI)
800 GOTO 860
810 T1=1/R-3/(8*Z)-15/(64*R*Z^2)-45/(512*Z^3)+315/(8192*R*Z^4)
820 T2=1/R+1/(8*Z)+9/(64*R*Z^2)+39/(512*Z^3)+75/(8192*R*Z^4)
830 D4=1+1/(4*R*Z)+1/(64*Z^2)-33/(256*R*Z^3)-1797/(8192*Z^4)
840 TR=T1/D4
850 TI=T2/D4
860 FFR=(.25*Z*(TR*(1-2*TI/Z)+TI*TR*2/Z))/((1-2*TI/Z)^2+(2*TR/Z)^2)
870 FI=(.25*Z*(TI*(1-2*TI/Z)-TR*TR*2/Z))/((1-2*TI/Z)^2+(2*TR/Z)^2)
```

```
880 VMOD=BR+4*GR/3
890 OMEG=F(I)*PI*2
900 AR=(KR*KR-KI*KI-HR*DR+HI*DI)/(VMOD*VMOD)
910 AI=(2*KR*KI-HI*DR-HR*DI)/(VMOD*VMOD)
920 ER=(HR*M/RHO+DR-2*KR*FRHO/RHO+(FFR*HI+FI*HR)*VISCF/(RHO*PERM*OMEG))/VMOD
930 EI=(HI*M/RHO+DI-2*KI*FRHO/RHO+(FI*HI-FFR*HR)*VISCF/(RHO*PERM*OMEG))/VMOD
940 CR=FRHO*FRHO/(RHO*RHO)-M/RHO-FI*VISCF/(RHO*PERM*OMEG)
950 CI=FFR*VISCF/(RHO*PERM*OMEG)
960 D2=2*(AR*AR+AI*AI)
970 RR=ER*ER-EI*EI-4*(AR*CR-AI*CI)
980 RI=2*EI*ER-4*(AI*CR+AR*CI)
990 RTMOD=(RR*RR+RI*RI)^.25
1000 IF ABS(RI/RR) > 1 THEN ANGL=PI/2-ABS(ATN(RR/RI)) ELSE ANGL=ABS(ATN(RI/RR))
1010 IF RR<0 THEN BATA=(PI-ANGL)/2 ELSE BATA=ANGL/2
1020 IF RI<0 THEN BATA=-BATA
1030 R1=RTMOD*COS(BATA)
1040 R2=RTMOD*SIN(BATA)
1050 D1=(ER+R1)*(ER+R1)+(EI+R2)*(EI+R2)
1060 X1=-2*(CR*(ER+R1)+CI*(EI+R2))/D1
1070 X2=(AR*(-ER-R1)-AI*(EI+R2))/D2
1080 Y1=2*(CI*(-ER-R1)+CR*(EI+R2))/D1
1090 Y2=(AI*(ER+R1)-AR*(EI+R2))/D2
1100 CC=OMEG*SQR(RHO/VMOD)
1110 IF ABS(Y1/X1)>1 THEN AA=PI/2-ABS(ATN(X1/Y1)) ELSE AA=ABS(ATN(Y1/X1))
1120 IF X1<0 THEN AA=(PI-AA)/2 ELSE AA=AA/2
1130 IF Y1<0 THEN AA=-AA
1140 IF ABS(Y2/X2)>1 THEN AB=PI/2-ABS(ATN(X2/Y2)) ELSE AB=ABS(ATN(Y2/X2))
1150 IF X2<0 THEN AB=(PI-AB)/2 ELSE AB=AB/2
1160 IF Y2<0 THEN AB=-AB
1170 P1=(X1*X1+Y1*Y1)^.25
1180 P2=(X2*X2+Y2*Y2)^.25
1190 V1=OMEG/(P1*COS(AA)*CC*100)
1200 A1=P1*SIN(AA)*CC*100
1210 V2=OMEG/(P2*COS(AB)*CC*100)
1220 A2=P2*SIN(AB)*CC*100
1230 Q1=V1*A1/F(I):Q2=V2*A2/F(I)
1240 E1=GR*M*OMEG*OMEG+GR*VISCF*OMEG*FI/PERM+GI*FFR*VISCF*OMEG/PERM
1250 E2=GI*M*OMEG*OMEG+GI*VISCF*OMEG*FI/PERM-GR*FFR*VISCF*OMEG/PERM
1260 XR=(RHO*M-FRHO*FRHO)*OMEG^4+VISCF*OMEG*FI/PERM*RHO*OMEG*OMEG
1270 XI=(-RHO*VISCF*FFR*OMEG^3)/PERM
1280 E1=E1/1E+15:E2=E2/1E+15:XR=XR/1E+15:XI=XI/1E+15
1290 X=(XR*E1+XI*E2)/(E1*E1+E2*E2)
1300 Y=(XI*E1-XR*E2)/(E1*E1+E2*E2)
1310 P=(X*X+Y*Y)^.25:THETA=ATN(Y/X)/2:GOTO 1320
1320 V3=OMEG/(P*COS(THETA)*100)
1330 A3=P*SIN(THETA)*100:Q3=V3*A3/F(I)
1340 LPRINT TAB(10) F(I) TAB(20):LPRINT USING FFF$;V1;:LPRINT USING
     FF$;-A1;-A1*8.686;:LPRINT USING FFFF$;-Q1;:LPRINT USING FFF$;V3;:
     LPRINT USING FF$;-A3;-A3*8.686;:LPRINT USING FFFF$;-Q3;
1350 LPRINT USING FFF$;V2;:LPRINT USING FFFF$;-Q2;:LPRINT USING FZ$;GDEC
1360 NEXT I
1370 END

4900 '******** SUBROUTINE TO INPUT PARAMETERS FOR COLE _ COLE MODEL **********
4910 PRINT "Typical values for the low frequency decrement are in the range
     of .01 to .04 for sands, sandstones, etc. and .07 to .10 for silts and
     stiff clay. Experiments on silts suggest values for amplitude factor,
     alpha, and peak frequency of .15, .2, & 10000, respectively.
```

```
5000 INPUT "LOW FREQUENCY DECREMENT";LOWDEC
5010 INPUT "AMPLITUDE FACTOR";AMFAC
5020 INPUT "ALPHA";ALPHA
5030 INPUT "FREQUENCY AT PEAK DECREMENT";PEAKFREQ
5040 LPRINT:LPRINT TAB(10) "PARAMETERS FOR VISCOELASTIC FRAME"
5050 LPRINT
5060 LPRINT TAB(10) "LOW FREQ LOG DECREMENT ---- ";LOWDEC
5070 LPRINT TAB(10) "AMPLITUDE FACTOR ---------- ";AMFAC
5080 LPRINT TAB(10) "ALPHA -------------------- ";ALPHA
5090 LPRINT TAB(10) "FREQ AT PEAK DECREMENT ---- ";PEAKFREQ
5100 RETURN
6000 IF G$ = "C" OR G$ = "c" THEN DEC(I) = GDEC:GOTO 6200
6005 TAU = 1/(2*PI*PEAKFREQ)
6010 MR1 = GR
6020 MI1 = LOWDEC*MR1/PI
6040 MR2 = MR1*AMFAC
6050 A1A = 1 - ALPHA
6060 OMEGA = F(I)*2*PI
6070 DENOM = 1 + 2*(OMEGA*TAU)^A1A*SIN(ALPHA*PI/2) + (OMEGA*TAU)^(2*A1A)
6080 MR3 = MR2 - (MR2*(1+(OMEGA*TAU)^A1A*SIN(ALPHA*PI/2)))/DENOM
6090 MI3 = MR2*(OMEGA*TAU)^A1A*COS(ALPHA*PI/2)/DENOM
6100 MR = MR1 + MR3
6110 MI = MI1 + MI3
6120 DEC(I) = MI/MR*PI
6200 RETURN
```

APPENDIX - B
CALCULATION OF REFLECTION COEFFICIENTS

In addition to the two vectors **u** and **U** (see Eq. 1.1) another vector quantity

$$\mathbf{w} = \beta(\mathbf{u} - \mathbf{U})$$

is useful in deriving expressions for the reflection coefficients. Using this vector the equations of motion (Eqs. 1.15 and 1.17) become

$$\mu \nabla^2 \mathbf{u} + (H - 2\mu)\nabla e - C\nabla\zeta = \rho \ddot{\mathbf{u}} - \rho_f \ddot{\mathbf{w}} \qquad (B.1)$$

$$C\nabla e - M\nabla\zeta = \rho_f \ddot{\mathbf{u}} - m\ddot{\mathbf{w}} - \eta/k\,\dot{\mathbf{w}}.$$

In solving Eqs. (B.1), it is convenient to express the vectors **u** and **w** in terms of scalar and vector potentials so that

$$\mathbf{u} = \nabla\Phi_s + curl\,\Psi_s \qquad (B.2)$$

$$\mathbf{w} = \nabla\Phi_f + curl\,\Psi_f.$$

Thus the volumetric strain of the skeletal frame, e, is given by

$$e = div\,\mathbf{u} = \nabla^2\Phi_s$$

and the increment of fluid content can be expressed as

$$\zeta = \nabla^2\Phi_f.$$

If we substitute (B.2) and apply the divergence operator to Eq. (B.1) the result is a pair of coupled equations that determine the scalar potentials Φ_s and Φ_f.

$$H\nabla^2\Phi_s - C\nabla^2\Phi_f = \rho\ddot{\Phi}_s - \rho_f\ddot{\Phi}_f \qquad (B.3)$$

$$C\nabla^2\Phi_s - M\nabla^2\Phi_f = \rho_f\ddot{\Phi}_s - m\ddot{\Phi}_f - \eta/k\dot{\Phi}_f.$$

Similarly, applying the curl operation to Eqs. (B.1) yields another pair of coupled equations for the vector potentials Ψ_s and Ψ_f

$$\mu\nabla^2\Psi_s = \rho\ddot{\Psi}_s - \rho_f\ddot{\Psi}_f \qquad (B.4)$$

$$\eta/k\dot{\Psi}_f = \rho_f\ddot{\Psi}_s - m\ddot{\Psi}_f.$$

Equations (B.3) admit solutions of the form

$$\Phi_s = A\exp[i(\omega t - \mathbf{k} \cdot \mathbf{r})] \qquad (B.5)$$

$$\Phi_f = B\exp[i(\omega t - \mathbf{k} \cdot \mathbf{r})]$$

where **k** is a complex vector

$$\mathbf{k} = \mathbf{k}_p + i\mathbf{k}_a.$$

$\mathbf{k}_p$ is a real vector that determines the direction of the normal to planes of equal phase, and $\mathbf{k}_a$ is a real vector that determines the direction of the normal to planes of equal amplitude.

When solutions of the form given by Eqs. (B.5) are substituted into Eqs. (B.3), we obtain a frequency equation identical to Eq. 1.21 with l given by

$$\mathbf{k} \cdot \mathbf{k} = l^2$$

In the case of homogeneous waves, the vectors $\mathbf{k}_p$ and $\mathbf{k}_a$ are in the same direction so that $\mathbf{k}$ can be simplified to a complex number rather than a complex vector. However, in the present case we are interested in inhomogeneous waves so that $\mathbf{k}_p$ and $\mathbf{k}_a$ are generally not in the same direction.

The frequency equation corresponding to Eqs. B.5 can be derived in a similar manner. With

$$\Psi_s = C \exp[i(\omega t - \mathbf{k} \cdot \mathbf{r})] \qquad\qquad (B.6)$$

$$\Psi_f = D \exp\{i(\omega t - \mathbf{k} \cdot \mathbf{r})]$$

the frequency equation given by Eq. 1.24 is obtained.

In the case of homogeneous plane waves, the real and imaginary parts of the complex wavenumber, $l = l_r + i l_i$, are completely determined by the roots of the frequency equations. However, when inhomogeneous waves are generated at an interface, $\mathbf{k}_s$ and $\mathbf{k}_p$ are variables that depend on conditions at the boundary. For example, if a homogeneous wave propagating in water is incident to a water-sediment interface, it is necessary that the projection of the real propagation vector onto the interface be equal to the corresponding projections of any waves that are generated at the interface (Snell's law). Thus

$$l_1 \sin\theta_1 = l_2 \sin\theta_2$$

where l_1 is the real wavenumber in water and θ_1 is the angle of incidence of the water wave. When an inhomogeneous wave is refracted into the sediment, l_2 is the modulus of the vector $\mathbf{k}$ and therefore will be a complex number. Thus $\sin\theta_2$ must also be complex so that the product $l_2 \sin\theta_2$ is real in order for the above equation to be satisfied.

In the special case of a homogeneous wave incident in water, the interface will be a plane of equal amplitude with spatial periodicity $2\pi/l\sin\theta_1$. Thus the planes of equal amplitude in the sediment will be parallel to the interface as shown in Fig. B.1

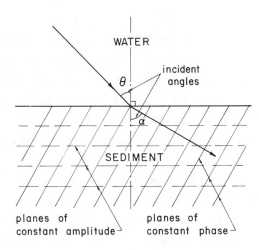

Fig. B.1. Orientation of planes of constant phase and constant amplitude (Stoll and Kan, 1983)

If the incoming wave in water is described by the potential

$$\Phi_i = A_i \exp[i(\omega t - l_1\sin\theta_1 x - l_1\cos\theta_1 z)] \qquad (B.7)$$

and a transmitted wave by the potential

$$\Phi_t = A_t \exp[i(\omega t - l_2\sin\theta_2 x - l_2\cos\theta_2 z)] \qquad (B.8)$$

where $l_2 = l_{2r} + il_{2i}$, then the complex angle θ_2 is defined by the expression

$$\sin\theta_2 = l_c/(l_{2r} + il_{2i})$$

and thus

$$\cos\theta_2 = \pm\{1 - [l_c^2/(l_{2r} + il_{2i})^2]\}^{1/2}$$

where $l_c = l_1\sin\theta_1$. Finally if we set

$$l_2\cos\theta_2 = k_{2r} + ik_{2i}$$

then Eq. B.8 takes the form

$$\Phi_t = A_t \exp[i(\omega t - l_c x - k_{2r} z)]\exp(k_{2i}z). \qquad (B.9)$$

Thus in the case of water over sediment, with a homogeneous plane wave incident in the water, the complex vector k is defined by

$$|\mathbf{k}_p| = (l_c^2 + k_{2r}^2)^{1/2}$$

$$\alpha = arg(\mathbf{k}_p) = \tan^{-1}(l_c/k_{2r}) \qquad (B.10)$$

$$|\mathbf{k}_a| = k_{2i}$$

$$arg(\mathbf{k}_a) = 0$$

Real angle α defines the direction of propagation of planes of constant phase so that $|\mathbf{k}_p|$ and $|\mathbf{k}_a|$ may be thought of as the equal phase and equal amplitude wavenumbers. In all real materials, $|arg(\mathbf{k}_p) - arg(\mathbf{k}_a)| < \pi/2$ so that α is less than 90° for all angles of incidence in the water. Thus there is no "critical" angle in the sense that the direction of phase propagation becomes parallel to the interface.

In order to solve the complete problem of reflection and refraction at the water sediment interface, a series of potentials of the form given by Eqs. B.7 - B.9 are used to represent the incident and reflected water waves and three refracted waves in the sediment as shown in Fig. B.2. The incident and reflected water waves have the displacement potentials

$$\Phi_i = A_i \exp[i(\omega t - k_w \cos\theta z - k_c x)] \qquad (B.11)$$

$$\Phi_r = A_r \exp[i(\omega t + k_w \cos\theta z - k_c x)] \qquad (B.12)$$

where $k_c = k_w \sin\theta$ and $k_w = \omega/v_w$. v_w is the phase velocity in water.

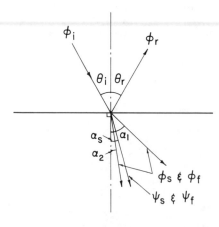

Fig. B.2. Incident, reflected and refracted potentials (Stoll and Kan, 1983).

In the sediment the scalar and vector potentials defined by Eqs.
B.2 are

$$\Phi_s = A_1 \exp[i(\omega t - k_{1p} \cos\alpha_1 z - k_c x)]\exp(k_{1a}z)$$

$$+ A_2 \exp[i(\omega t - k_{2p} \cos\alpha_2 z - k_c x)]\exp(k_{2a}z) \qquad (B.13)$$

$$\Phi_f = B_1 \exp[i(\omega t - k_{1p} \cos\alpha_1 z - k_c x)]\exp(k_{2a}z)$$

$$+ B_2 \exp[i(\omega t - k_{2p} \cos\alpha_2 z - k_c x)]\exp(k_{2a}z) \qquad (B.14)$$

$$\Psi_s = C \exp[i(\omega t - k_{sp} \cos\alpha_s z - k_c x)]\exp(k_{sa}z) \qquad (B.15)$$

$$\Psi_f = D \exp[i(\omega t - k_{sp} \cos\alpha_s z - k_c x)]\exp(k_{sa}z) \qquad (B.16)$$

where k_{1p} and k_{2p} are the equal phase wavenumbers for the first
and second kind of dilatational waves and k_{1a} and k_{2a} are the equal
amplitude wavenumbers. k_{sp} and k_{sa} are the equal phase and equal
amplitude wavenumbers for the shear wave.

When Eqs. B.13 - B.16 are substituted into Eqs. B.3 and B.4,
the following relationships result between the complex amplitudes
A_1, A_2, B_1, B_2, C, and D

$$m_1 = m_{1r} + im_{1i} = B_1/A_1 = \overline{H}[\overline{H}l_1^2/(\rho\omega^2) - 1]/[C\overline{H}l_1^2/(\rho\omega^2) - \rho_f \overline{H}/\rho] \qquad (B.17)$$

$$m_2 = m_{2r} + im_{2i} = B_2/A_2 = \overline{H}[\overline{H}l_2^2/(\rho\omega^2) - 1]/[C\overline{H}l_2^2/(\rho\omega^2) - \rho_f \overline{H}/\rho] \qquad (B.18)$$

$$Q = Q_r + iQ_i = D/C = [1 - \overline{\mu}l_s^2/(\rho\omega^2)]/(\rho_f/\rho) \qquad (B.19)$$

where l_1, l_2 and l_s are the complex wave numbers for the dilatational
waves of the first and second kind and the shear wave in the
sediment, respectively. When Eqs. B.11 - B.19 are combined, there
remain four unknown complex amplitudes, A_1, A_2, A_r, and C assuming
A_i is specified. These amplitudes may be found by considering
the following boundary conditions at the interface z=0:
1. For continuity of fluid movement in and out of the skeletal
frame in a direction normal to the water-sediment interface

$$U_z' = u_z(1-\beta) + \beta U_z = u_z - w_z \qquad (B.20)$$

where U_z' is the net vertical displacement in the water. In terms
of displacement potentials this requires that

$$\frac{\partial \Phi_i}{\partial z} + \frac{\partial \Phi_r}{\partial z} = \frac{\partial \Phi_s}{\partial z} + \frac{\partial \Psi_s}{\partial x} - \frac{\partial \Phi_f}{\partial z} - \frac{\partial \Psi_f}{\partial x} \qquad (B.21)$$

2. For equilibrium of normal traction

$$\overline{H}e - 2\overline{\mu}\epsilon_{xx} - \overline{C}\zeta = -p_w \qquad (B.21)$$

or

$$\overline{H}\left(\frac{\partial^2\Phi_s}{\partial x^2}+\frac{\partial^2\Phi_s}{\partial z^2}\right)-2\overline{\mu}\left(\frac{\partial^2\Phi_s}{\partial x^2}-\frac{\partial^2\Psi_s}{\partial x\partial z}\right)-\overline{C}\left(\frac{\partial^2\Phi_f}{\partial x^2}+\frac{\partial^2\Phi_f}{\partial z^2}\right)=\rho_f\frac{\partial^2(\Phi_i+\Phi_r)}{\partial t^2}\qquad (B.23)$$

3.   For equilibrium of fluid pressure

$$\overline{M}\zeta-\overline{C}e=p_w$$

or

$$\overline{M}\left(\frac{\partial^2\Phi_f}{\partial x^2}+\frac{\partial^2\Phi_f}{\partial z^2}\right)-\overline{C}\left(\frac{\partial^2\Phi_s}{\partial x^2}+\frac{\partial^2\Phi_s}{\partial z^2}\right)=-\rho_f\frac{\partial^2(\Phi_i+\Phi_r)}{\partial t^2}\qquad (B.25)$$

4.   For equilibrium of tangential traction

$$\sigma_{zx}=0 \qquad (B.26)$$

or

$$2\overline{\mu}\frac{\partial^2\Phi_s}{\partial x\partial z}-\overline{\mu}\left(\frac{\partial^2\Psi_s}{\partial z^2}-\frac{\partial^2\Psi_s}{\partial x^2}\right)=0 \qquad (B.27)$$

By combining the Eqs. B.11 - B.27 we may derive a set of four linear complex equations or eight linear real equations which determine the four unknown complex amplitudes $A_r=A_{rr}+iA_{ri}$, $A_1=A_{1r}+iA_{1i}$, $A_2=A_{2r}+iA_{2i}$ and $C=C_r+iC_i$. The eight real equations are given by

$$\begin{pmatrix} C_{11} & C_{12} & C_{13} & \cdots & C_{18} \\ C_{21} & C_{22} & C_{23} & \cdots & C_{28} \\ \cdot & \cdot & \cdot & & \cdot \\ \cdot & \cdot & \cdot & & \cdot \\ \cdot & \cdot & \cdot & & \cdot \\ \cdot & \cdot & \cdot & & \cdot \\ \cdot & \cdot & \cdot & & \cdot \\ C_{81} & C_{82} & C_{83} & \cdots & C_{88} \end{pmatrix} \times \begin{pmatrix} A_{rr} \\ A_{ri} \\ A_{1r} \\ A_{1i} \\ A_{2r} \\ A_{2i} \\ C_r \\ C_i \end{pmatrix} = \begin{pmatrix} Y_1 \\ Y_2 \\ \cdot \\ \cdot \\ \cdot \\ \cdot \\ \cdot \\ Y_8 \end{pmatrix} \qquad (B.28)$$

The components of {C} and {Y} are given below. Eq. B.28 was used to calculate the values for the reflection curves shown in Chap. 2.

$C_{11} = 0$

$C_{12} = k_w \cos \theta_i$

$C_{13} = k_{1a} - m_{1i} k_{1p} \cos \alpha_1 - k_{1a} m_{1r}$

$C_{14} = k_{1p} \cos \alpha_1 - m_{1r} k_{1p} \cos \alpha_1 + k_{1a} m_{1i}$

$C_{15} = k_{2a} - m_{2i} k_{2p} \cos \alpha_2 - k_{2a} m_{2r}$

$C_{16} = k_{2p} \cos \alpha_2 - m_{2r} k_{2p} \cos \alpha_2 + k_{2a} m_{2i}$.

$C_{17} = -k_c Q_i$

$C_{18} = k_c - k_c Q_r$

$Y_1 = 0$

$C_{21} = -k_w \cos \theta_i$

$C_{22} = 0$

$C_{23} = -k_{1p} \cos \alpha_1 + m_{1r} k_{1p} \cos \alpha_1 - k_{1a} m_{1i}$

$C_{24} = k_{1a} - m_{1i} k_{1p} \cos \alpha_1 - m_{1r} k_{1a}$

$C_{25} = -k_{2p} \cos \alpha_2 + m_{2r} k_{2p} \cos \alpha_2 - k_{2a} m_{2i}$

$C_{26} = k_{2a} - m_{2i} k_{2p} \cos \alpha_2 - m_{2r} k_{2a}$

$C_{27} = -k_c + k_c Q_r$

$C_{28} = -k_c Q_i$

$Y_2 = -k_w \cos \theta_i A_i$

$$C_{31} = \rho_w \omega^2$$

$$C_{32} = 0$$

$$C_{33} = 2k_c^2 G_r + H_r(k_{1a}^2 - k_{1p}^2 \cos^2 \alpha_1 - k_c^2) + 2H_i k_{1a} k_{1p} \cos \alpha_1$$
$$- (C_r m_{1r} - C_i m_{1i})(k_{1a}^2 - k_{1p}^2 \cos^2 \alpha_1 - k_c^2) - 2k_{1a} k_{1p} \cos \alpha_1 (C_r m_{1i} + C_i m_{1r})$$

$$C_{34} = -2k_c^2 G_i + 2k_{1a} k_{1p} \cos \alpha_1 H_r - (k_{1a}^2 - k_{1p}^2 \cos^2 \alpha_1 - k_c^2) H_i$$
$$- 2k_{1a} k_{1p} \cos \alpha_1 (C_r m_{1r} - C_i m_{1i}) + (C_r m_{1i} + C_i m_{1r})(k_{1a}^2 - k_{1p}^2 \cos^2 \alpha_1 - k_c^2)$$

$$C_{35} = 2k_c^2 G_r + H_r(k_{2a}^2 - k_{2p}^2 \cos^2 \alpha_2 - k_c^2) + 2H_i k_{2a} k_{2p} \cos \alpha_2$$
$$- (C_r m_{2r} - C_i m_{2i})(k_{2a}^2 - k_{2p}^2 \cos^2 \alpha_2 - k_c^2) - 2k_{2a} k_{2p} \cos \alpha_2 (C_r m_{2i} + C_i m_{2r})$$

$$C_{36} = -2k_c^2 G_i + 2k_{2a} k_{2p} \cos \alpha_2 H_r - (k_{2a}^2 - k_{2p}^2 \cos^2 \alpha_2 - K_c^2) H_i$$
$$- k_{2a} k_{2p} \cos \alpha_2 (C_r m_{2r} - C_i m_{2i}) + (C_r m_{2i} + C_i m_{2r})(k_{2a}^2 - k_{2p}^2 \cos^2 \alpha_2 - k_c^2)$$

$$C_{37} = -2k_{sp} \cos \alpha_s k_c G_r + 2k_c k_{sa} G_i$$

$$C_{38} = 2k_{sp} \cos \alpha_s k_c G_i + 2k_c k_{sa} G_r$$

$$Y_3 = -\rho_w \omega^2 A_i$$

$$C_{41} = 0$$

$$C_{42} = \rho_w \omega^2$$

$$C_{43} = 2k_c^2 G_i + (k_{1a}^2 - k_{1p}^2 \cos^2 \alpha_1 - k_c^2) H_i - 2k_{1a} k_{1p} \cos \alpha_1 H_r$$
$$- (C_r m_{1i} + C_i m_{1r})(k_{1a}^2 - k_{1p}^2 \cos^2 \alpha_1 - k_c^2) + 2k_{1a} k_{1p} \cos \alpha_1 (C_r m_{1r} - C_i m_{1i})$$

$$C_{44} = 2k_c^2 G_r + 2k_{1a} k_{1p} \cos \alpha_1 H_i + (k_{1a}^2 - k_{1p}^2 \cos^2 \alpha_1 - k_c^2) H_r$$
$$- 2(C_r m_{1i} + C_i m_{1r}) k_{1a} k_{1a} \cos \alpha_1 - (C_r m_{1r} - C_i m_{1i})(k_{1a}^2 - k_{1p}^2 \cos^2 \alpha_1 - k_c^2)$$

$$C_{45} = 2k_c^2 G_i + (k_{2a}^2 - k_{2p}^2 \cos^2 \alpha_2 - k_c^2) H_i - 2k_{2a} k_{2p} \cos \alpha_2 H_r$$

$$- (C_r m_{2i} + C_i m_{2r})(k_{2a}^2 - k_{2p}^2 \cos^2 \alpha_2 - k_c^2) + 2k_{2a} k_{2p} \cos \alpha_2 (C_r m_{2r} - C_i m_{2i})$$

$$C_{46} = 2k_c^2 G_r + 2k_{2a} k_{2p} \cos \alpha_2 H_i + (k_{2a}^2 - k_{2p}^2 \cos^2 \alpha_2 - k_c^2) H_r$$

$$- 2(C_r m_{2i} + C_i m_{2r})k_{2a} k_{2p} \cos \alpha_2 + (C_r m_{2r} - C_i m_{2i})(k_{2a}^2 - k_{2p}^2 \cos^2 \alpha_2 - k_c^2)$$

$$C_{47} = -2k_{sp} \cos \alpha_s k_c G_i - 2k_c k_{sa} G_r$$

$$C_{48} = -2k_{sp} \cos \alpha_s k_c G_r + 2k_c k_{sa} G_r$$

$$Y_4 = 0$$

- -

$$C_{51} = -\rho_w \omega^2$$

$$C_{52} = 0$$

$$C_{53} = (M_r m_{1r} - M_i m_{1i})(k_{1a}^2 - k_{1p}^2 \cos^2 \alpha_1 - k_c^2)$$

$$+ 2k_{1a} k_{1p} \cos \alpha_1 (M_r m_{1i} + M_i m_{1r}) - C_r (k_{1a}^2 - k_{1p}^2 \cos^2 \alpha_1 - k_c^2) - 2C_i k_{1a} k_{1p} \cos \alpha_1$$

$$C_{54} = 2k_{1a} k_{1p} \cos \alpha_1 (M_r m_{1r} - M_i m_{1i}) - (k_{1a}^2 - k_{1p}^2 \cos^2 \alpha_1 - k_c^2)(M_r m_{1i} - M_i m_{1r})$$

$$- 2k_{1a} k_{1p} \cos \alpha_1 C_r + (k_{1a}^2 - k_{1p}^2 \cos^2 \alpha_1 - k_c^2) C_i$$

$$C_{55} = (M_r m_{2r} - M_i m_2 i)(k_{2a}^2 \cos^2 \alpha_2 - k_c^2)$$

$$+ 2k_{2a} k_{2p} \cos \alpha_2 (M_r m_{2i} + M_i m_{2r}) - C_r (k_{2a}^2 - k_{2p}^2 \cos^2 \alpha_2 - k_c^2) - 2C_i k_{2a} k_{2p} \cos \alpha_2$$

$$C_{56} = 2k_{2a} k_{2p} \cos \alpha_s (M_r m_{2i} - M_i m_{2p}) - (k_{2a}^2 - k_{2p}^2 \cos^2 \alpha_2 - k_c^2)(M_r m_{2i} + M_i m_{2r})$$

$$- 2k_{2a} k_{2p} \cos \alpha_2 C_r + (k_{2a}^2 - k_{2p}^2 \cos^2 \alpha_2 - k_c^2) C_i$$

$$C_{57} = 0$$

$$C_{58} = 0$$

$$Y_5 = \rho_w \omega^2$$

$$C_{61} = 0$$

$$C_{62} = -\rho_w \omega^2$$

$$C_{63} = (M_r m_{1i} + M_i m_{1r})(k_{1a}^2 k_{1p}^2 \cos^2 \alpha_1 - k_c^2)$$

$$- 2k_{1a} k_{1p} \cos \alpha_1 (M_r m_{1r} - M_i m_{1i}) - (k_{1a}^2 - k_{1p}^2 \cos^2 \alpha_1 - k_c^2) C_i + 2k_{1a} k_{1p} \cos \alpha_1 C_r$$

$$C_{64} = 2(M_r m_{1i} + M_i m_{1r}) k_{1a} k_{1p} \cos \alpha_1 + (M_r m_{1r} - M_i m_{1i})(k_{1a}^2 - k_{1p}^2 \cos^2 \alpha_1 - k_c^2)$$

$$- 2k_{1a} k_{1p} \cos \alpha_1 C_i - (k_{1a}^2 - k_{1p}^2 \cos^2 \alpha_1 - k_c^2) C_r$$

$$C_{65} = (M_r m_{2i} + M_i m_{2r})(k_{2a}^2 k_{2p}^2 \cos^2 \alpha_2 - k_c^2)$$

$$- 2k_{2a} k_{2p} \cos \alpha_2 (M_r m_{2r} - M_i m_{2i}) - (k_{2a}^2 - k_{2p}^2 \cos^2 \alpha_2 - k_c^2) C_i + 2k_{2a} k_{2p} \cos \alpha_2 C_r$$

$$C_{66} = 2(M_r m_{2i} + M_i m_{2r}) k_{2a} k_{2p} \cos \alpha_2$$

$$+ (M_r m_{2r} - M_i m_{2i})(k_{2a}^2 - k_{2p}^2 \cos^2 \alpha_2 - k_c^2) - 2k_{2a} k_{2p} \cos \alpha_2 C_i - (k_{2a}^2 - k_{2p}^2 \cos^2 \alpha_2 - k_c^2) C_r$$

$$C_{67} = 0$$

$$C_{68} = 0$$

$$Y_6 = 0$$

$$C_{71} = 0$$

$$C_{72} = 0$$

$$C_{73} = 2G_i k_c k_{1a} - 2G_r k_c k_{1p} \cos \alpha_1$$

$$C_{74} = 2G_i k_c k_{1p} \cos \alpha_1 + 2G_r k_c k_{1a}$$

$$C_{75} = 2G_i k_c k_{2a} - 2G_r k_c k_{2p} \cos \alpha_2$$

$$C_{76} = 2G_i k_c k_{2a} \cos \alpha_2 + 2G_r k_c k_{2a}$$

$$C_{77} = G_r(-k_c^2 - k_{sa}^2 + k_{sp}^2 \cos^2 \alpha_s) - G_i(2k_{sa} k_{sp} \cos \alpha_s)$$

$$C_{78} = -G_i(-k_c^2 - k_{sa}^2 + k_{sp}^2 \cos^2 \alpha_s) + 2k_{sa} k_{sp} \cos \alpha_s G_r$$

$$Y_7 = 0$$

$$C_{81} = 0$$

$$C_{82} = 0$$

$$C_{83} = -2k_c k_{1a} G_r - 2k_c k_{1p} \cos\alpha_1 G_i$$

$$C_{84} = 2G_i k_c k_{1a} - 2k_c k_{1p} \cos\alpha_1 G_r$$

$$C_{85} = -2k_c k_{2a} G_r - 2k_c k_{2p} \cos\alpha_2 G_i$$

$$C_{86} = 2G_i k_c k_{2a} - 2k_c k_{2p} \cos\alpha_2 G_r$$

$$C_{87} = 2k_{sa} k_{sp} \cos\alpha_s G_r + G_i(-k_c^2 - k_{sa}^2 + k_{sp}^2 \cos^2\alpha_s)$$

$$C_{88} = (-k_c^2 - k_{sa}^2 + k_{sp}^2 \cos^2\alpha_s)G_r - 2k_{sa} k_{sp} \cos\alpha_s G_i$$

$$Y_8 = 0$$

**EQUATIONS TO DETERMINE MODE SHAPES IN TORSIONAL
RESONANT COLUMN EXPERIMENTS**

Solutions of the equations of motion given by Eq. 1.24 of
the form

$$\omega = A\exp[i(\omega t \pm lz)] \qquad\qquad (C.1)$$

$$\theta = B\exp[i(\omega t \pm lz)]$$

may be used to construct solutions for the case of torsional
vibrations of a cylindrical column of sediment of finite length.
By taking a linear combination of positive and negative going
waves described by Eq. C.1, we obtain the equation of a standing
wave that may be matched to the boundary conditions at the top
and bottom of a specimen. Utilizing a cylindrical coordinate
system r, θ, and z, with z the axis of the cylinder, in the first
approximation we expect no radial or axial displacements and
assume that the tangential displacement u_θ is of the form

$$u_\theta = r\overline{\omega}_z(z,t)$$

so that

$$\overline{\omega}_z = \partial u_\theta / \partial r = A_1\exp[i(\omega t - lz)] + A_2\exp[i(\omega t + lz)] \qquad (C.2)$$

where $\overline{\omega}_z$ is the rotation of the crossection perpendicular to the
axis of the specimen. If the specimen is fixed at the bottom,
$z = 0$, then $\overline{\omega}_z(0,t) = 0$ which requires that $A_2 = -A_1$. Thus $\overline{\omega}_z$ may be
written as

$$\overline{\omega}_z = A\sin(lz)e^{i\omega t}, \qquad A = -2iA_1 \qquad\qquad (C.3)$$

In experiments where a rigid mass is attached to the top of
the specimen, the equation of motion of the mass will be

$$T = -I\frac{\partial^2[\overline{\omega}_z(h,t)]}{\partial t^2} + Me^{i\omega t} \qquad\qquad (C.4)$$

where I is the mass polar moment of inertia of the mass including
the driver coil and any transducers that are attached, M is the
amplitude of applied torque, and T is the resultant torque
transmitted to the top of the specimen. This torque may be related
to the rotation of the top face of the specimen by considering
the relationship between traction on the top face and the shear

strain. Since we are considering harmonic motion, this is most conveniently accomplished by using the notion of a complex modulus to express the relationship of stress to strain,

$$\tau_{\theta z} = (\mu_r + i\mu_i)\gamma_{\theta z}.$$

$\tau_{\theta z}$ is the tangential component of traction on the upper face of the specimen, $\gamma_{\theta z}$ is equal to $\partial u_\theta / \partial z = r \partial \overline{\omega}_z / \partial z$, and $\mu = \mu_r + i\mu_i$ is the overall complex shear modulus at a given frequency. Integrating over the area of the top face gives

$$T = J(\mu_r + i\mu_i)\frac{\partial \overline{\omega}_z}{\partial z}, \qquad\qquad (C.5)$$

where J is the area polar moment of inertia of the top face. When μ_i is small compared to μ_r, the propagation constant l, determined from Eq. 1.24 is related to μ by the relations $l_i/l_r = \mu_i/2\mu_r$ and $\mu_r = \rho(\omega/l_r)^2$. Combining Eqs. C.3, C.4, and C.5 we obtain the following equation which may be used to determine the value of A

$$A[lGJ\cos(lh) - I\omega^2 \sin(lh)] = M \qquad\qquad (C.5)$$

The real part of $\overline{\omega}_z$ from Eq. C.3 gives the response of the system to a driving torque $M\cos\omega t$ and the imaginary part to $M\sin\omega t$.

Fig. C.1 shows the kind of response described by Eq. C.3. The ordinates of this figure, labeled "resonance factor", represent the ratio of the rotation of the top of the specimen to the rotation for an equivalent static torque and the abcissa is the dimensionless frequency $F = 2\omega h/\pi V$ where h is the length of the specimen and V is phase velocity. To determine shear wave velocity from a resonant column experiment, it is necessary to match the resonant frequncies predicted by Eq. C.3 with experimentally observed values. Since μ is a complex function of frequency, this is most easily accomplished by performing iterative calculations on a computer. A typical program used for this purpose is given by Hardin and Music (1965).

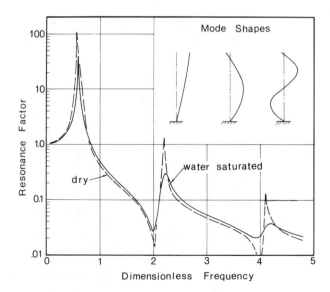

Fig. C.1.   Steady state response of a specimen with fixed
base and a top mass (Stoll, 1979).

Index